Information Retrieval in Chemistry and Chemical Patent Law

Information Retrieval in Chemistry and Chemical Patent Law

ENCYCLOPEDIA REPRINT SERIES

Editor: Martin Grayson

A WILEY-INTERSCIENCE PUBLICATION

JOHN WILEY & SONS

NEW YORK · CHICHESTER · BRISBANE · TORONTO · SINGAPORE

Library of Congress Cataloging in Publication Data

Main entry under title:

Information retrieval in chemistry and chemical patent law.

  (Encyclopedia reprint series; v. 3)
  Articles reprinted from the Kirk-Othmer
encyclopedia of chemical technology. 3rd ed.
  "A Wiley-Interscience publication." Includes
bibliographical references and index.
  1. Information storage and retrieval systems—
Chemistry. 2. Information storage and retrieval systems—
Chemistry—Patents. I. Grayson, Martin. II. Encyclopedia
of chemical technology. 3rd ed.

Z699.5.C5153   1983   025'.0654   82-24727
ISBN 0-471-89057-X

Printed in the United States of America

10 9 8 7 6 5 4 3 2 1

CONTENTS

EDITORIAL STAFF

CONTRIBUTORS

Margaret H. Graham, *Exxon Research and Engineering Company, Linden, New Jersey,* Information retrieval
Alexis B. Lamy, *Essochem Europe, Inc., Diegem, Belgium,* Information retrieval
Barbara Lawrence, *Exxon Corporation, Linden, New Jersey,* Information retrieval
John W. Lotz, *IFI/Plenum Data Company, Wilmington, Delaware,* Patents, literature
Lorraine Y. Stroumtsos, *Exxon Research and Engineering Company, Linden, New Jersey,* Information retrieval

PREFACE

This volume in the Encyclopedia Reprint Series is addressed to the needs of information and patent law professionals and students preparing for careers in the chemical and allied industries. The articles in this volume are taken from the internationally recognized, authoritative *Kirk-Othmer Encyclopedia of Chemical Technology*, third edition. The authors are information professionals from industry who are experts in their field. The articles were reviewed by competent specialists and carefully edited for clarity, accuracy, and readability by the Wiley editorial staff. The full text and extensive bibliographies, charts, and figures of the original articles have been reproduced here unchanged. Introductory information from the Encyclopedia concerning Chemical Abstracts Registry Numbers, nomenclature, SI units and conversion factors, and related information has been provided for the reader as a further guide to those concerned with chemical technology and especially the information aspects of this field. It is expected that this volume will serve as a useful ready reference for the information specialist as well as supplementary course material for teaching professionals and their students.

In addition to thorough coverage of the print sources of chemical and patent information, these articles provide detailed guidance to the electronic data bases and alerting services that constitute an increasingly vital aspect of the search tools of the information specialist in the chemical industry. Moreover, the Information Retrieval article reviews the key sources of business information for industry from an international point of view. This aspect alone more than justifies the low cost of this useful volume.

M. GRAYSON

NOTE ON CHEMICAL ABSTRACTS SERVICE REGISTRY NUMBERS AND NOMENCLATURE

Chemical Abstracts Service (CAS) Registry Numbers are unique numerical identifiers assigned to substances recorded in the CAS Registry System. They appear in brackets in the *Chemical Abstracts* (CA) substance and formula indexes following the names of compounds. A single compound may have many synonyms in the chemical literature. A simple compound like phenethylamine can be named β-phenylethylamine or, as in *Chemical Abstracts*, benzeneethanamine. The usefulness of the *Encyclopedia* depends on accessibility through the most common correct name of a substance. Because of this diversity in nomenclature careful attention has been given the problem in order to assist the reader as much as possible, especially in locating the systematic CA index name by means of the Registry Number. For this purpose, the reader may refer to the CAS Registry Handbook-Number Section which lists in numerical order the Registry Number with the *Chemical Abstracts* index name and the molecular formula; eg, **458-88-8,** Piperidine, 2-propyl-, (*S*)-, $C_8H_{17}N$; in the *Encyclopedia* this compound would be found under its common name, coniine [*458-88-8*]. The Registry Number is a valuable link for the reader in retrieving additional published information on substances and also as a point of access for such on-line data bases as Chemline, Medline, and Toxline.

In all cases, the CAS Registry Numbers have been given for title compounds in articles and for all compounds in the index. All specific substances indexed in *Chemical Abstracts* since 1965 are included in the CAS Registry System as are a large number of substances derived from a variety of reference works. The CAS Registry System identifies a substance on the basis of an unambiguous computer-language description of its molecular structure including stereochemical detail. The Registry Number is a machine-checkable number (like a Social Security number) assigned in sequential order to each substance as it enters the registry system. The value of the number lies in the fact that it is a concise and unique means of substance identification, which is independent of, and therefore bridges, many systems of chemical nomenclature. For polymers, one Registry Number is used for the entire family; eg, polyoxyethylene (20) sorbitan monolaurate has the same number as all of its polyoxyethylene homologues.

Registry numbers for each substance will be provided in the third edition cumulative index and appear as well in the annual indexes (eg, Alkaloids shows the Registry Number of all alkaloids (title compounds) in a table in the article as well, but the intermediates have their Registry Numbers shown only in the index). Articles such as Analytical methods, Batteries and electric cells, Chemurgy, Distillation, Economic evaluation, and Fluid mechanics have no Registry Numbers in the text.

Cross-references are inserted in the index for many common names and for some systematic names. Trademark names appear in the index. Names that are incorrect, misleading or ambiguous are avoided. Formulas are given very frequently in the text to help in identifying compounds. The spelling and form used, even for industrial names, follow American chemical usage, but not always the usage of *Chemical Abstracts* (eg, *coniine* is used instead of *(S)-2-propylpiperidine*, *aniline* instead of *benzenamine*, and *acrylic acid* instead of *2-propenoic acid*).

There are variations in representation of rings in different disciplines. The dye industry does not designate aromaticity or double bonds in rings. All double bonds and aromaticity are shown in the *Encyclopedia* as a matter of course. For example, tetralin has an aromatic ring and a saturated ring and its structure appears in the

Encyclopedia with its common name, Registry Number enclosed in brackets, and parenthetical CA index name, ie, tetralin, [*119-64-2*] (1,2,3,4-tetrahydronaphthalene). With names and structural formulas, and especially with CAS Registry Numbers the aim is to help the reader have a concise means of substance identification.

CONVERSION FACTORS, ABBREVIATIONS, AND UNIT SYMBOLS

SI Units (Adopted 1960)

A new system of measurement, the International System of Units (abbreviated SI), is being implemented throughout the world. This system is a modernized version of the MKSA (meter, kilogram, second, ampere) system, and its details are published and controlled by an international treaty organization (The International Bureau of Weights and Measures) (1).

SI units are divided into three classes:

BASE UNITS

length	meter[†] (m)
mass[‡]	kilogram (kg)
time	second (s)
electric current	ampere (A)
thermodynamic temperature[§]	kelvin (K)
amount of substance	mole (mol)
luminous intensity	candela (cd)

[†] The spellings "metre" and "litre" are preferred by ASTM; however "-er" are used in the Encyclopedia.

[‡] "Weight" is the commonly used term for "mass."

[§] Wide use is made of "Celsius temperature" (t) defined by

$$t = T - T_0$$

where T is the thermodynamic temperature, expressed in kelvins, and $T_0 = 273.15$ K by definition. A temperature interval may be expressed in degrees Celsius as well as in kelvins.

SUPPLEMENTARY UNITS

plane angle   radian (rad)
solid angle   steradian (sr)

DERIVED UNITS AND OTHER ACCEPTABLE UNITS

These units are formed by combining base units, supplementary units, and other derived units (2–4). Those derived units having special names and symbols are marked with an asterisk in the list below:

Quantity	Unit	Symbol	Acceptable equivalent
*absorbed dose	gray	Gy	J/kg
acceleration	meter per second squared	m/s^2	
*activity (of ionizing radiation source)	becquerel	Bq	1/s
area	square kilometer	km^2	
	square hectometer	hm^2	ha (hectare)
	square meter	m^2	
*capacitance	farad	F	C/V
concentration (of amount of substance)	mole per cubic meter	mol/m^3	
*conductance	siemens	S	A/V
current density	ampere per square meter	A/m^2	
density, mass density	kilogram per cubic meter	kg/m^3	g/L; mg/cm^3
dipole moment (quantity)	coulomb meter	C·m	
*electric charge, quantity of electricity	coulomb	C	A·s
electric charge density	coulomb per cubic meter	C/m^3	
electric field strength	volt per meter	V/m	
electric flux density	coulomb per square meter	C/m^2	
*electric potential, potential difference, electromotive force	volt	V	W/A
*electric resistance	ohm	Ω	V/A
*energy, work, quantity of heat	megajoule	MJ	
	kilojoule	kJ	
	joule	J	N·m
	electron volt[†]	eV[†]	
	kilowatt-hour[†]	kW·h[†]	

[†] This non-SI unit is recognized by the CIPM as having to be retained because of practical importance or use in specialized fields (1).

Quantity	Unit	Symbol	Acceptable equivalent
energy density	joule per cubic meter	J/m^3	
*force	kilonewton	kN	
	newton	N	$kg{\cdot}m/s^2$
*frequency	megahertz	MHz	
	hertz	Hz	$1/s$
heat capacity, entropy	joule per kelvin	J/K	
heat capacity (specific), specific entropy	joule per kilogram kelvin	$J/(kg{\cdot}K)$	
heat transfer coefficient	watt per square meter kelvin	$W/(m^2{\cdot}K)$	
*illuminance	lux	lx	lm/m^2
*inductance	henry	H	Wb/A
linear density	kilogram per meter	kg/m	
luminance	candela per square meter	cd/m^2	
*luminous flux	lumen	lm	$cd{\cdot}sr$
magnetic field strength	ampere per meter	A/m	
*magnetic flux	weber	Wb	$V{\cdot}s$
*magnetic flux density	tesla	T	Wb/m^2
molar energy	joule per mole	J/mol	
molar entropy, molar heat capacity	joule per mole kelvin	$J/(mol{\cdot}K)$	
moment of force, torque	newton meter	$N{\cdot}m$	
momentum	kilogram meter per second	$kg{\cdot}m/s$	
permeability	henry per meter	H/m	
permittivity	farad per meter	F/m	
*power, heat flow rate, radiant flux	kilowatt	kW	
	watt	W	J/s
power density, heat flux density, irradiance	watt per square meter	W/m^2	
*pressure, stress	megapascal	MPa	
	kilopascal	kPa	
	pascal	Pa	N/m^2
sound level	decibel	dB	
specific energy	joule per kilogram	J/kg	
specific volume	cubic meter per kilogram	m^3/kg	
surface tension	newton per meter	N/m	
thermal conductivity	watt per meter kelvin	$W/(m{\cdot}K)$	
velocity	meter per second	m/s	
	kilometer per hour	km/h	
viscosity, dynamic	pascal second	$Pa{\cdot}s$	
	millipascal second	$mPa{\cdot}s$	
viscosity, kinematic	square meter per second	m^2/s	

Quantity	Unit	Symbol	Acceptable equivalent
	square millimeter per second	mm^2/s	
volume	cubic meter	m^3	
	cubic decimeter	dm^3	L(liter) (5)
	cubic centimeter	cm^3	mL
wave number	1 per meter	m^{-1}	
	1 per centimeter	cm^{-1}	

In addition, there are 16 prefixes used to indicate order of magnitude, as follows:

Multiplication factor	Prefix	Symbol	Note
10^{18}	exa	E	
10^{15}	peta	P	
10^{12}	tera	T	
10^9	giga	G	
10^6	mega	M	
10^3	kilo	k	
10^2	hecto	h[a]	[a] Although hecto, deka, deci, and centi
10	deka	da[a]	are SI prefixes, their use should be
10^{-1}	deci	d[a]	avoided except for SI unit-mul-
10^{-2}	centi	c[a]	tiples for area and volume and
10^{-3}	milli	m	nontechnical use of centimeter,
10^{-6}	micro	μ	as for body and clothing
10^{-9}	nano	n	measurement.
10^{-12}	pico	p	
10^{-15}	femto	f	
10^{-18}	atto	a	

For a complete description of SI and its use the reader is referred to ASTM E 380 (4) and the article Units and Conversion Factors which will appear in a later volume of the *Encyclopedia*.

A representative list of conversion factors from non-SI to SI units is presented herewith. Factors are given to four significant figures. Exact relationships are followed by a dagger. A more complete list is given in ASTM E 380-79(4) and ANSI Z210.1-1976 (6).

Conversion Factors to SI Units

To convert from	To	Multiply by
acre	square meter (m^2)	4.047×10^3
angstrom	meter (m)	1.0×10^{-10}†
are	square meter (m^2)	1.0×10^2†
astronomical unit	meter (m)	1.496×10^{11}
atmosphere	pascal (Pa)	1.013×10^5
bar	pascal (Pa)	1.0×10^5†
barn	square meter (m^2)	1.0×10^{-28}†

† Exact.

To convert from	To	Multiply by
barrel (42 U.S. liquid gallons)	cubic meter (m^3)	0.1590
Bohr magneton (μ_β)	J/T	9.274×10^{-24}
Btu (International Table)	joule (J)	1.055×10^3
Btu (mean)	joule (J)	1.056×10^3
Btu (thermochemical)	joule (J)	1.054×10^3
bushel	cubic meter (m^3)	3.524×10^{-2}
calorie (International Table)	joule (J)	4.187
calorie (mean)	joule (J)	4.190
calorie (thermochemical)	joule (J)	4.184[†]
centipoise	pascal second (Pa·s)	1.0×10^{-3}[†]
centistoke	square millimeter per second (mm^2/s)	1.0[†]
cfm (cubic foot per minute)	cubic meter per second (m^3/s)	4.72×10^{-4}
cubic inch	cubic meter (m^3)	1.639×10^{-5}
cubic foot	cubic meter (m^3)	2.832×10^{-2}
cubic yard	cubic meter (m^3)	0.7646
curie	becquerel (Bq)	3.70×10^{10}[†]
debye	coulomb·meter (C·m)	3.336×10^{-30}
degree (angle)	radian (rad)	1.745×10^{-2}
denier (international)	kilogram per meter (kg/m)	1.111×10^{-7}
	tex[‡]	0.1111
dram (apothecaries')	kilogram (kg)	3.888×10^{-3}
dram (avoirdupois)	kilogram (kg)	1.772×10^{-3}
dram (U.S. fluid)	cubic meter (m^3)	3.697×10^{-6}
dyne	newton (N)	1.0×10^{-5}[†]
dyne/cm	newton per meter (N/m)	1.0×10^{-3}[†]
electron volt	joule (J)	1.602×10^{-19}
erg	joule (J)	1.0×10^{-7}[†]
fathom	meter (m)	1.829
fluid ounce (U.S.)	cubic meter (m^3)	2.957×10^{-5}
foot	meter (m)	0.3048[†]
footcandle	lux (lx)	10.76
furlong	meter (m)	2.012×10^{-2}
gal	meter per second squared (m/s^2)	1.0×10^{-2}[†]
gallon (U.S. dry)	cubic meter (m^3)	4.405×10^{-3}
gallon (U.S. liquid)	cubic meter (m^3)	3.785×10^{-3}
gallon per minute (gpm)	cubic meter per second (m^3/s)	6.308×10^{-5}
	cubic meter per hour (m^3/h)	0.2271
gauss	tesla (T)	1.0×10^{-4}
gilbert	ampere (A)	0.7958
gill (U.S.)	cubic meter (m^3)	1.183×10^{-4}
grad	radian	1.571×10^{-2}
grain	kilogram (kg)	6.480×10^{-5}
gram force per denier	newton per tex (N/tex)	8.826×10^{-2}

[†] Exact.

[‡] See footnote on p. xiv.

To convert from	To	Multiply by
hectare	square meter (m^2)	$1.0 \times 10^{4\dagger}$
horsepower (550 ft·lbf/s)	watt (W)	7.457×10^2
horsepower (boiler)	watt (W)	9.810×10^3
horsepower (electric)	watt (W)	$7.46 \times 10^{2\dagger}$
hundredweight (long)	kilogram (kg)	50.80
hundredweight (short)	kilogram (kg)	45.36
inch	meter (m)	$2.54 \times 10^{-2\dagger}$
inch of mercury (32°F)	pascal (Pa)	3.386×10^3
inch of water (39.2°F)	pascal (Pa)	2.491×10^2
kilogram force	newton (N)	9.807
kilowatt hour	megajoule (MJ)	$3.6^\dagger$
kip	newton (N)	4.48×10^3
knot (international)	meter per second (m/s)	0.5144
lambert	candela per square meter (cd/m^2)	3.183×10^3
league (British nautical)	meter (m)	5.559×10^3
league (statute)	meter (m)	4.828×10^3
light year	meter (m)	9.461×10^{15}
liter (for fluids only)	cubic meter (m^3)	$1.0 \times 10^{-3\dagger}$
maxwell	weber (Wb)	$1.0 \times 10^{-8\dagger}$
micron	meter (m)	$1.0 \times 10^{-6\dagger}$
mil	meter (m)	$2.54 \times 10^{-5\dagger}$
mile (statute)	meter (m)	1.609×10^3
mile (U.S. nautical)	meter (m)	$1.852 \times 10^{3\dagger}$
mile per hour	meter per second (m/s)	0.4470
millibar	pascal (Pa)	1.0×10^2
millimeter of mercury (0°C)	pascal (Pa)	$1.333 \times 10^{2\dagger}$
minute (angular)	radian	2.909×10^{-4}
myriagram	kilogram (kg)	10
myriameter	kilometer (km)	10
oersted	ampere per meter (A/m)	79.58
ounce (avoirdupois)	kilogram (kg)	2.835×10^{-2}
ounce (troy)	kilogram (kg)	3.110×10^{-2}
ounce (U.S. fluid)	cubic meter (m^3)	2.957×10^{-5}
ounce-force	newton (N)	0.2780
peck (U.S.)	cubic meter (m^3)	8.810×10^{-3}
pennyweight	kilogram (kg)	1.555×10^{-3}
pint (U.S. dry)	cubic meter (m^3)	5.506×10^{-4}
pint (U.S. liquid)	cubic meter (m^3)	4.732×10^{-4}
poise (absolute viscosity)	pascal second (Pa·s)	$0.10^\dagger$
pound (avoirdupois)	kilogram (kg)	0.4536
pound (troy)	kilogram (kg)	0.3732
poundal	newton (N)	0.1383
pound-force	newton (N)	4.448
pound per square inch (psi)	pascal (Pa)	6.895×10^3
quart (U.S. dry)	cubic meter (m^3)	1.101×10^{-3}

† Exact.

To convert from	To	Multiply by
quart (U.S. liquid)	cubic meter (m^3)	9.464×10^{-4}
quintal	kilogram (kg)	$1.0 \times 10^{2\dagger}$
rad	gray (Gy)	$1.0 \times 10^{-2\dagger}$
rod	meter (m)	5.029
roentgen	coulomb per kilogram (C/kg)	2.58×10^{-4}
second (angle)	radian (rad)	4.848×10^{-6}
section	square meter (m^2)	2.590×10^6
slug	kilogram (kg)	14.59
spherical candle power	lumen (lm)	12.57
square inch	square meter (m^2)	6.452×10^{-4}
square foot	square meter (m^2)	9.290×10^{-2}
square mile	square meter (m^2)	2.590×10^6
square yard	square meter (m^2)	0.8361
stere	cubic meter (m^3)	$1.0^\dagger$
stokes (kinematic viscosity)	square meter per second (m^2/s)	$1.0 \times 10^{-4\dagger}$
tex	kilogram per meter (kg/m)	$1.0 \times 10^{-6\dagger}$
ton (long, 2240 pounds)	kilogram (kg)	1.016×10^3
ton (metric)	kilogram (kg)	$1.0 \times 10^{3\dagger}$
ton (short, 2000 pounds)	kilogram (kg)	9.072×10^2
torr	pascal (Pa)	1.333×10^2
unit pole	weber (Wb)	1.257×10^{-7}
yard	meter (m)	$0.9144^\dagger$

Abbreviations and Unit Symbols

Following is a list of commonly used abbreviations and unit symbols appropriate for use in the *Encyclopedia*. In general they agree with those listed in *American National Standard Abbreviations for Use on Drawings and in Text (ANSI Y1.1)* (6) and *American National Standard Letter Symbols for Units in Science and Technology (ANSI Y10)* (6). Also included is a list of acronyms for a number of private and government organizations as well as common industrial solvents, polymers, and other chemicals.

Rules for Writing Unit Symbols (4):

1. Unit symbols should be printed in upright letters (roman) regardless of the type style used in the surrounding text.

2. Unit symbols are unaltered in the plural.

3. Unit symbols are not followed by a period except when used as the end of a sentence.

4. Letter unit symbols are generally written in lower-case (eg, cd for candela) unless the unit name has been derived from a proper name, in which case the first letter of the symbol is capitalized (W,Pa). Prefix and unit symbols retain their prescribed form regardless of the surrounding typography.

5. In the complete expression for a quantity, a space should be left between the numerical value and the unit symbol. For example, write 2.37 lm, *not* 2.37lm, and 35 mm, *not* 35mm. When the quantity is used in an adjectival sense, a hyphen is often used, for example, 35-mm film. *Exception:* No space is left between the numerical value and the symbols for degree, minute, and second of plane angle, and degree Celsius.

6. No space is used between the prefix and unit symbols (eg, kg).

7. Symbols, not abbreviations, should be used for units. For example, use "A," not "amp," for ampere.

8. When multiplying unit symbols, use a raised dot:

$$N{\cdot}m \text{ for newton meter}$$

In the case of W·h, the dot may be omitted, thus:

$$Wh$$

An exception to this practice is made for computer printouts, automatic typewriter work, etc, where the raised dot is not possible, and a dot on the line may be used.

9. When dividing unit symbols use one of the following forms:

$$m/s \; or \; m{\cdot}s^{-1} \; or \; \frac{m}{s}$$

In no case should more than one slash be used in the same expression unless parentheses are inserted to avoid ambiguity. For example, write:

$$J/(mol{\cdot}K) \; or \; J{\cdot}mol^{-1} \cdot K^{-1} \; or \; (J/mol)/K$$

but *not*

$$J/mol/K$$

10. Do not mix symbols and unit names in the same expression. Write:

$$joules \; per \; kilogram \; or \; J/kg \; or \; J{\cdot}kg^{-1}$$

but *not*

$$joules/kilogram \; nor \; joules/kg \; nor \; joules{\cdot}kg^{-1}$$

ABBREVIATIONS AND UNITS

A	ampere	AIME	American Institute of Mining, Metallurgical, and Petroleum Engineers
A	anion (eg, H*A*); mass number		
a	atto (prefix for 10^{-18})		
AATCC	American Association of Textile Chemists and Colorists	AIP	American Institute of Physics
ABS	acrylonitrile–butadiene–styrene	AISI	American Iron and Steel Institute
abs	absolute	alc	alcohol(ic)
ac	alternating current, *n*.	Alk	alkyl
a-c	alternating current, *adj*.	alk	alkaline (not alkali)
ac-	alicyclic	amt	amount
acac	acetylacetonate	amu	atomic mass unit
ACGIH	American Conference of Governmental Industrial Hygienists	ANSI	American National Standards Institute
		AO	atomic orbital
ACS	American Chemical Society	AOAC	Association of Official Analytical Chemists
AGA	American Gas Association	AOCS	American Oil Chemist's Society
Ah	ampere hour		
AIChE	American Institute of Chemical Engineers	APHA	American Public Health Association

API American Petroleum Institute
aq aqueous
Ar aryl
ar- aromatic
as- asymmetric(al)
ASH-
 RAE American Society of Heating, Refrigerating, and Air Conditioning Engineers
ASM American Society for Metals
ASME American Society of Mechanical Engineers
ASTM American Society for Testing and Materials
at no. atomic number
at wt atomic weight
av(g) average
AWS American Welding Society
b bonding orbital
bbl barrel
bcc body-centered cubic
BCT body-centered tetragonal
Bé Baumé
BET Brunauer-Emmett-Teller (adsorption equation)
bid twice daily
Boc *t*-butyloxycarbonyl
BOD biochemical (biological) oxygen demand
bp boiling point
Bq becquerel
C coulomb
°C degree Celsius
C- denoting attachment to carbon
c centi (prefix for 10^{-2})
c critical
ca circa (approximately)
cd candela; current density; circular dichroism
CFR Code of Federal Regulations
cgs centimeter–gram–second
CI Color Index
cis- isomer in which substituted groups are on same side of double bond between C atoms
cl carload

cm centimeter
cmil circular mil
cmpd compound
CNS central nervous system
CoA coenzyme A
COD chemical oxygen demand
coml commercial(ly)
cp chemically pure
cph close-packed hexagonal
CPSC Consumer Product Safety Commission
cryst crystalline
cub cubic
D Debye
D- denoting configurational relationship
d differential operator
d- dextro-, dextrorotatory
da deka (prefix for 10^1)
dB decibel
dc direct current, *n.*
d-c direct current, *adj.*
dec decompose
detd determined
detn determination
Di didymium, a mixture of all lanthanons
dia diameter
dil dilute
DIN Deutsche Industrie Normen
dl-; DL- racemic
DMA dimethylacetamide
DMF dimethylformamide
DMG dimethyl glyoxime
DMSO dimethyl sulfoxide
DOD Department of Defense
DOE Department of Energy
DOT Department of Transportation
DP degree of polymerization
dp dew point
DPH diamond pyramid hardness
dstl(d) distill(ed)
dta differential thermal analysis
(*E*)- entgegen; opposed
ϵ dielectric constant (unitless number)
e electron

ECU	electrochemical unit	GRAS	Generally Recognized as Safe
ed.	edited, edition, editor	grd	ground
ED	effective dose	Gy	gray
EDTA	ethylenediaminetetraacetic acid	H	henry
		h	hour; hecto (prefix for 10^2)
emf	electromotive force	ha	hectare
emu	electromagnetic unit	HB	Brinell hardness number
en	ethylene diamine	Hb	hemoglobin
eng	engineering	hcp	hexagonal close-packed
EPA	Environmental Protection Agency	hex	hexagonal
		HK	Knoop hardness number
epr	electron paramagnetic resonance	HRC	Rockwell hardness (C scale)
		HV	Vickers hardness number
eq.	equation	hyd	hydrated, hydrous
esp	especially	hyg	hygroscopic
esr	electron-spin resonance	Hz	hertz
est(d)	estimate(d)	^{i}(eg, Pr^i)	iso (eg, isopropyl)
estn	estimation	i-	inactive (eg, i-methionine)
esu	electrostatic unit	IACS	International Annealed Copper Standard
exp	experiment, experimental		
ext(d)	extract(ed)	ibp	initial boiling point
F	farad (capacitance)	IC	inhibitory concentration
F	faraday (96,487 C)	ICC	Interstate Commerce Commission
f	femto (prefix for 10^{-15})		
FAO	Food and Agriculture Organization (United Nations)	ICT	International Critical Table
		ID	inside diameter; infective dose
		ip	intraperitoneal
fcc	face-centered cubic	IPS	iron pipe size
FDA	Food and Drug Administration	IPTS	International Practical Temperature Scale (NBS)
FEA	Federal Energy Administration		
		ir	infrared
fob	free on board	IRLG	Interagency Regulatory Liaison Group
fp	freezing point		
FPC	Federal Power Commission	ISO	International Organization for Standardization
FRB	Federal Reserve Board		
frz	freezing	IU	International Unit
G	giga (prefix for 10^9)	IUPAC	International Union of Pure and Applied Chemistry
G	gravitational constant = 6.67×10^{11} N·m^2/kg^2		
		IV	iodine value
g	gram	iv	intravenous
(g)	gas, only as in H_2O(g)	J	joule
g	gravitational acceleration	K	kelvin
gem-	geminal	k	kilo (prefix for 10^3)
glc	gas-liquid chromatography	kg	kilogram
g-mol wt; gmw	gram-molecular weight	L	denoting configurational relationship
GNP	gross national product	L	liter (for fluids only)(5)
gpc	gel-permeation chromatography	l-	$levo$-, levorotatory
		(l)	liquid, only as in NH_3(l)

LC_{50}	conc lethal to 50% of the animals tested	mxt	mixture
LCAO	linear combination of atomic orbitals	μ	micro (prefix for 10^{-6})
		N	newton (force)
LCD	liquid crystal display	N	normal (concentration); neutron number
lcl	less than carload lots		
LD_{50}	dose lethal to 50% of the animals tested	N-	denoting attachment to nitrogen
LED	light-emitting diode	n (as n_D^{20})	index of refraction (for 20°C and sodium light)
liq	liquid		
lm	lumen	^{n} (as Bu^n), n-	normal (straight-chain structure)
ln	logarithm (natural)		
LNG	liquefied natural gas	n	neutron
log	logarithm (common)	n	nano (prefix for 10^9)
LPG	liquefied petroleum gas	na	not available
ltl	less than truckload lots	NAS	National Academy of Sciences
lx	lux		
M	mega (prefix for 10^6); metal (as in $M\Lambda$)	NASA	National Aeronautics and Space Administration
M	molar; actual mass	nat	natural
$\overline{M}_w$	weight-average mol wt	NBS	National Bureau of Standards
$\overline{M}_n$	number-average mol wt		
m	meter; milli (prefix for 10^{-3})	neg	negative
m	molal	NF	*National Formulary*
m-	meta	NIH	National Institutes of Health
max	maximum		
MCA	Chemical Manufacturers' Association (was Manufacturing Chemists Association)	NIOSH	National Institute of Occupational Safety and Health
		nmr	nuclear magnetic resonance
MEK	methyl ethyl ketone	NND	New and Nonofficial Drugs (AMA)
meq	milliequivalent		
mfd	manufactured	no.	number
mfg	manufacturing	NOI- (BN)	not otherwise indexed (by name)
mfr	manufacturer		
MIBC	methyl isobutyl carbinol	NOS	not otherwise specified
MIBK	methyl isobutyl ketone	nqr	nuclear quadruple resonance
MIC	minimum inhibiting concentration	NRC	Nuclear Regulatory Commission; National Research Council
min	minute; minimum		
mL	milliliter	NRI	New Ring Index
MLD	minimum lethal dose	NSF	National Science Foundation
MO	molecular orbital	NTA	nitrilotriacetic acid
mo	month	NTP	normal temperature and pressure (25°C and 101.3 kPa or 1 atm)
mol	mole		
mol wt	molecular weight		
mp	melting point	NTSB	National Transportation Safety Board
MR	molar refraction		
ms	mass spectrum	O-	denoting attachment to oxygen

o-	ortho	ref.	reference
OD	outside diameter	rf	radio frequency, *n*.
OPEC	Organization of Petroleum Exporting Countries	r-f	radio frequency, *adj*.
		rh	relative humidity
		RI	Ring Index
o-phen	*o*-phenanthridine	rms	root-mean square
OSHA	Occupational Safety and Health Administration	rpm	rotations per minute
		rps	revolutions per second
owf	on weight of fiber	RT	room temperature
Ω	ohm	^{s} (eg, Bu^s); *sec*-	secondary (eg, secondary butyl)
P	peta (prefix for 10^{15})		
p	pico (prefix for 10^{-12})		
p-	para	S	siemens
p	proton	(*S*)-	sinister (counterclockwise configuration)
p.	page		
Pa	pascal (pressure)	*S*-	denoting attachment to sulfur
pd	potential difference		
pH	negative logarithm of the effective hydrogen ion concentration	*s*-	symmetric(al)
		s	second
		(s)	solid, only as in $H_2O(s)$
phr	parts per hundred of resin (rubber)	SAE	Society of Automotive Engineers
p-i-n	positive-intrinsic-negative	SAN	styrene–acrylonitrile
pmr	proton magnetic resonance	sat(d)	saturate(d)
p-n	positive-negative	satn	saturation
po	per os (oral)	SBS	styrene–butadiene–styrene
POP	polyoxypropylene	sc	subcutaneous
pos	positive	SCF	self-consistent field; standard cubic feet
pp.	pages		
ppb	parts per billion (10^9)	Sch	Schultz number
ppm	parts per million (10^6)	SFs	Saybolt Furol seconds
ppmv	parts per million by volume	SI	Le Système International
ppmwt	parts per million by weight		d'Unités (International
PPO	poly(phenyl oxide)		System of Units)
ppt(d)	precipitate(d)	sl sol	slightly soluble
pptn	precipitation	sol	soluble
Pr (no.)	foreign prototype (number)	soln	solution
pt	point; part	soly	solubility
PVC	poly(vinyl chloride)	sp	specific; species
pwd	powder	sp gr	specific gravity
py	pyridine	sr	steradian
qv	quod vide (which see)	std	standard
R	univalent hydrocarbon radical	STP	standard temperature and pressure (0°C and 101.3 kPa)
(*R*)-	rectus (clockwise configuration)		
		sub	sublime(s)
r	precision of data	SUs	Saybolt Universal seconds
rad	radian; radius		
rds	rate determining step	syn	synthetic

^{t} (eg, Bu^t), t-, tert-	tertiary (eg, tertiary butyl)	Twad	Twaddell
		UL	Underwriters' Laboratory
		USDA	United States Department of Agriculture
T	tera (prefix for 10^{12}); tesla (magnetic flux density)	USP	*United States Pharmacopeia*
		uv	ultraviolet
t	metric ton (tonne); temperature	V	volt (emf)
		var	variable
TAPPI	Technical Association of the Pulp and Paper Industry	*vic-*	vicinal
		vol	volume (not volatile)
tex	tex (linear density)	vs	versus
T_g	glass-transition temperature	v sol	very soluble
tga	thermogravimetric analysis	W	watt
THF	tetrahydrofuran	Wb	Weber
tlc	thin layer chromatography	Wh	watt hour
TLV	threshold limit value	WHO	World Health Organization (United Nations)
trans-	isomer in which substituted groups are on opposite sides of double bond between C atoms		
		wk	week
		yr	year
TSCA	Toxic Substance Control Act	(Z)-	zusammen; together; atomic number
TWA	time-weighted average		

Non-SI (Unacceptable and Obsolete) Units		*Use*
Å	angstrom	nm
at	atmosphere, technical	Pa
atm	atmosphere, standard	Pa
b	barn	cm^2
bar†	bar	Pa
bbl	barrel	m^3
bhp	brake horsepower	W
Btu	British thermal unit	J
bu	bushel	m^3; L
cal	calorie	J
cfm	cubic foot per minute	m^3/s
Ci	curie	Bq
cSt	centistokes	mm^2/s
c/s	cycle per second	Hz
cu	cubic	exponential form
D	debye	C·m
den	denier	tex
dr	dram	kg
dyn	dyne	N
dyn/cm	dyne per centimeter	mN/m
erg	erg	J
eu	entropy unit	J/K
°F	degree Fahrenheit	°C; K
fc	footcandle	lx
fl	footlambert	lx
fl oz	fluid ounce	m^3; L
ft	foot	m
ft·lbf	foot pound-force	J

† Do not use bar (10^5Pa) or millibar (10^2Pa) because they are not SI units, and are accepted internationally only for a limited time in special fields because of existing usage.

Non-SI (*Unacceptable and Obsolete*) Units		Use
gf den	gram-force per denier	N/tex
G	gauss	T
Gal	gal	m/s^2
gal	gallon	m^3; L
Gb	gilbert	A
gpm	gallon per minute	(m^3/s); (m^3/h)
gr	grain	kg
hp	horsepower	W
ihp	indicated horsepower	W
in.	inch	m
in. Hg	inch of mercury	Pa
in. H_2O	inch of water	Pa
in.-lbf	inch pound-force	J
kcal	kilogram-calorie	J
kgf	kilogram-force	N
kilo	for kilogram	kg
L	lambert	lx
lb	pound	kg
lbf	pound-force	N
mho	mho	S
mi	mile	m
MM	million	M
mm Hg	millimeter of mercury	Pa
mμ	millimicron	nm
mph	miles per hour	km/h
μ	micron	μm
Oe	oersted	A/m
oz	ounce	kg
ozf	ounce-force	N
η	poise	Pa·s
P	poise	Pa·s
ph	phot	lx
psi	pounds-force per square inch	Pa
psia	pounds-force per square inch absolute	Pa
psig	pounds-force per square inch gauge	Pa
qt	quart	m^3; L
°R	degree Rankine	K
rd	rad	Gy
sb	stilb	lx
SCF	standard cubic foot	m^3
sq	square	exponential form
thm	therm	J
yd	yard	m

BIBLIOGRAPHY

1. The International Bureau of Weights and Measures, BIPM (Parc de Saint-Cloud, France) is described on page 22 of Ref. 4. This bureau operates under the exclusive supervision of the International Committee of Weights and Measures (CIPM).
2. *Metric Editorial Guide (ANMC-78-1)* 3rd ed., American National Metric Council, 1625 Massachusetts Ave. N.W., Washington, D.C. 20036, 1978.
3. *SI Units and Recommendations for the Use of Their Multiples and of Certain Other Units (ISO 1000-1981)*, American National Standards Institute, 1430 Broadway, New York, N. Y. 10018, 1981.
4. Based on *ASTM E 380-82 (Standard for Metric Practice)*, American Society for Testing and Materials, 1916 Race Street, Philadelphia, Pa. 19103, 1982.
5. *Fed. Regist.*, Dec. 10, 1976 (41 FR 36414).
6. For ANSI address, see Ref. 3.

R. P. LUKENS
American Society for Testing and Materials

INFORMATION RETRIEVAL

The information explosion is more than a catch phrase to describe the proliferation of technical information in recent years. For many chemists, it marks a destruction, a detonation if you will, of the traditional ways of finding chemical facts with a resultant information fallout, the unexpected, often bewildering products of this sudden surge. Information is more abundant and more specialized than ever before; but chemists, amid this vast wealth of information, are often unable to obtain easily the particular facts they need. This article is for those individuals; it identifies and describes some of the main resources currently available and the methods of information retrieval as of January 1980. Certain sections also are of interest to information and business professionals. The first section gives an overview of information retrieval: the technology that has developed to handle the wealth of scientific material, its advantages and disadvantages, the consequences for chemists and the chemical industry, and the prospects for the future. The second section describes the particular characteristics of technical information and outlines the productive routes and useful resources for chemical fact-finding; it reviews the main abstracting and indexing services, handbooks, physical data compilations, encyclopedias and treatises, main computerized data bases, and on-line vendors of chemical information. The discussion of business information in the third section shows that technical–economic, business, and management literature differs from the highly structured technical literature and that it varies according to political, social, and legal environments.

DEVELOPMENTS IN INFORMATION-RETRIEVAL TECHNOLOGY

To discuss the advances in information retrieval requires familiarity with the terms used for the sources of searchable information. Primary sources represent either new information, new interpretations of old knowledge, or new compilations of known information. They include books, journals, and patents. Secondary sources serve as locators for primary information—they are the organizers and the condensers of primary literature. These include encyclopedias, dictionaries, and handbooks. Chemical Abstracts Service (CAS) is an example of an abstracting and indexing service that organizes the primary periodical literature. CAS has emerged as the single, most important secondary source of information for chemists and, through international agreements, has indeed become responsive to an international community. Tertiary sources direct the user to both the primary and secondary sources; among tertiary sources are guides and directories.

For many years, chemical indexing and abstracting services, in processing the large number of publications in the field of chemistry, provided a way to keep track of developments in particular fields. Thus they enabled chemists to reduce the time spent on literature searching. However, the chemical literature grew so rapidly that the chemist no longer could rely on indexes and abstracts. Also, the interdisciplinary nature of the work being done made it difficult to locate pertinent information that might have appeared in another subject category. Computerized retrieval of information pointed a way to satisfy these needs.

Searching for Information. The computer has become indispensable in information retrieval. There is, however, a wealth of material that predates computer-generated information. In the 1960s computers generated indexes to the literature and, in specialized systems, identified structures. These indexes were produced as paper products for manual searching. Extensive use of the computer to actually store the indexes for manipulation really started about 1970. The computer offers the ability to retrieve current information in ways never before possible, but until technology develops ways of providing access to the accumulated knowledge of the past, the traditional retrieval techniques are still needed.

The computer is an essential tool in the management of information as well. Chemical-information systems with their associated hardware and software deal with the collection, storage, and interpretation of chemical data. The hardware of these systems includes the computer itself and the physical equipment involved in the storage and processing of data. The software is the particular collection of computer programs that enable the computer to perform particular tasks. Together they provide the technology to handle large quantities of chemical data and to allow timely and inexpensive access to information. To understand the level to which this technology has developed and its affect on the working chemist, it is necessary to know the kinds of information available for retrieval and the physical forms that they take.

In the late 1960s the computer-readable tapes that were used in the production of printed publications of the main abstracting and indexing services began to be used for alerting purposes. This was the beginning of a significant change in information retrieval. These tapes were manipulated further to produce a variety of products. They became searchable in batch-mode; ie, they could be queried, and then scanned by the computer in one operation. Then, on-line systems were developed. In on-line systems, data is transmitted directly between a computer and remote terminals and is immediately processed by the computer. These systems provided a way to search the tapes interactively. The person at the terminal, in direct communication with the computer, could refine the query during the scanning process. The development of these systems led to a proliferation of machine-readable files called data bases.

Because purchasing and maintaining these data bases for in-house use was expensive, companies and search services emerged that collected data bases and sold access to them either for a fee based on use or by subscription. Several of these systems are accessible by dialing directly or by dialing one of the commercial data transmission networks such as TYMNET (TYMESHARE), TELENET, TELEX, or TWX. Once connected to the retrieval system, the user can look at the contents of that search service's data bases. This process of examining data bases at remote locations using a computer terminal is called on-line searching.

Several types of files are searchable on line. Bibliographic data bases provide references to source material. To identify relevant source material, the user chooses appropriate keywords and phrases, links them together using the words AND, OR, and NOT (Boolean logic), and interactively refines the query while scanning the results. Relevant references can be printed immediately on line, or they can be printed off line, ie, not printed during the direct communication with the computer, but afterward, at the remote computer center. Nonbibliographic or numerical data bases allow the user to retrieve specific data directly and offer the additional capabilities of adding and manipulating data for analysis. On-line dictionary and vocabulary files, useful for specific information, also guide the user to formulas, synonyms, particular spellings of words, and particular terms for optimum retrieval of information from other on-line data bases.

Automated systems offer two modes for information retrieval: retrospective searching, by which the user identifies material on a given subject; and current awareness, by which the user identifies the most recent material on a given subject to keep abreast of the latest developments. This activity, called selective dissemination of information (SDI), is a continual searching service.

Access to Documents. The availability of original articles is an increasing problem. Because retrieval systems readily provide references to international and interdisciplinary source material, libraries often receive requests for full documents from obscure journals or periodicals to which they do not subscribe. Furthermore, documents that are available in libraries are requested and photocopied more often.

Photocopying practices and the distribution of full copies of documents are now affected by a new United States copyright law. This law discusses in detail what individuals may copy for themselves and what an information center, library, or research center may copy for its clientele. Details of the new law are reviewed in ref. 1 (see Trademarks and copyrights). Other countries are considering similar laws.

Manipulating Chemical Data for Retrieval. Retrieving chemical information requires manipulation of textual material and chemical structures. Analysis of text involves extracting data and concepts from a document and expressing them in words or phrases. The bibliographic data elements by which a document is stored and retrieved include its title, author(s), subject, journal sources, date, and reference number. Keywords and descriptors identify the contents of the document and, unlike strictly bibliographic information, require intellectual effort to be useful. Comprehensive, in-depth indexing requires much intellectual effort but also yields the most useful tool for identifying relevant material. Technology is developing to have computers do more of the indexing work.

There have been many efforts to devise systems to represent the nature of chemical substances including systematic nomenclature and stereochemical drawings. Because structural diagrams are extremely difficult to convey in written text, systematic names have been developed. These names are unique textual representations of chemical structures, but they can be cumbersome and difficult to prepare correctly because of the number of rules that govern the naming process. Chemists revert to structural diagrams to define the chemical structures of compounds unequivocally. The most commonly used systems for the retrieval of structures are topological systems and linear-notation systems. The latter use codes to represent chemical fragments and the linking of atoms in the molecule; examples are the Wiswesser Line Notation and the IUPAC (International Union of Pure and Applied Chemistry)/Dyson Notation system, which parallels chemical nomenclature. Topological systems use connection tables or matrix representations of the topology of the chemical structures.

Registry Numbers. Chemical Abstracts Service's Registry Numbers serve as a useful bridge between textual and structural information. A Registry Number is a computer-checkable serial number assigned to a chemical compound by CAS in sequential order when the compound initially enters the registry system. This system, which originated in 1965, has registered over five million substances and additional compounds are being registered at the rate of over three hundred fifty thousand per year. A Registry Number has no chemical significance, but it is the unique identifier for a particular compound that links various textual identifiers of the substance (ie, nomenclature) with the molecular structure.

Registry Numbers for a particular substance can be found in the CAS *Registry Handbook,* or in the CAS *Chemical Substance Index, Formula, Index,* or *General Index.* They may be found also in chemical-dictionary files. Registry Numbers have appeared regularly in *Chemical Abstracts* volume indexes since 1972 and are an integral part of the CA SEARCH data base. The Registry Number appears as a data element in an ever increasing number of on-line chemical files and technical journals, and it is even required in reporting to the EPA. In the *Encyclopedia,* Registry Numbers have been given for title compounds in articles and for all compounds in the index as a concise means of substance identification. The use of Registry Numbers as a link between on-line bibliographic files is a refined retrieval technique not previously available.

Handling Proprietary Information.   In today's high technology environment, knowledge of the surrounding technology and some means for communicating that technology are essential; information is an integral part of the utilization of technology. This applies also to proprietary information, that body of knowledge collected and held privately by particular organizations. Just as published information is publicly available, a company's information must be available to those within that company who need it. The emphasis, however, must be on the transfer of information rather than on the transfer of paper. The process of transferring proprietary information is cyclic and involves several steps: (*1*) information generation, the technical activity that gives rise to data that is or will be of value to others; (*2*) information preparation, the making of a permanent record of the data; (*3*) information processing, the collection, indexing, storage, and distribution of the data; (*4*) information retrieval, the search for specific, requested data; (*5*) information analysis, the organization of data to improve its applicability; and (*6*) information application, the technical activity that arises from the transfer of data and that, in turn, leads to the generation of new data. These six processes are interdependent. The objective in handling proprietary information is to make the step between the generation and application of information as transparent as possible to the user and to eliminate the barriers (eg, location, organizational structure) between the generator of information and the user who wants to apply what was generated.

Advanced information and office equipment technology has evolved to meet these objectives. Historically, proprietary information was stored in internal files of some type; these files employed a classified arrangement ranging from very simple author files to very complex and elaborate hierarchical breakdowns of subjects. Now computer-produced indexes eliminate the need for manually listing the places where documents are filed. And because information needed in the course of business is not based just in one document, the need to identify multiple documents, to extract particular data, and to use merged information is served by improved technology.

Text processors, systems for capturing information in digital form when it is initially typed, are replacing typewriters. They permit information to be revised, stored, and manipulated readily. Although they were initially sold as improved typewriters, they are now being developed to accept and store information in digital form for processing and retrieval.

Telecommunications between machines are providing new services and options. Letters and reports typed in one location can be printed at remote locations. Electronic mail provides immediate delivery of messages without the use of a messenger. Printing-on-demand allows information stored in digital form to be printed as needed in a variety of ways.

All of these technical developments affect the handling of proprietary information, and work going on right now will affect it further. For example, the video disk currently used in entertainment, advertising, and publishing may be applied in the field of information storage and retrieval (see Recording disks). Another technology that bears watching is the voice-activated computer. In the current technology, most terminals have typewriter keyboards for entering information, but the voice-activated computer eliminates keystroking, opening the door to the office of the future in which a computer terminal on a desk top is as common as the telephone is today. This is indeed a projection into the future, but in view of the pace of technological development, it is worthy of consideration.

Consequences of Advanced Information Technology. There exists today an identifiable information industry. The ability to identify and obtain information through the use of computers is so powerful and has grown so fast that its effect is being felt throughout the world. The growth of an information industry based on repackaging the products of abstracting and indexing organizations is a new phenomenon. The repackaging of proprietary information within industrial organizations is occurring in a similar fashion and, of necessity, is causing the growth of information centers in large industries. Smaller organizations that cannot afford to have in-house capabilities use an information broker to supply their needs and may also use the expertise of the information consultant to organize their private collections. That information is a marketable commodity, with all its attendant professionals, is a new concept.

Sophisticated information technology often requires the use of a trained specialist. Because research and development is expensive, information must be used to avoid duplication of effort and to keep scientists abreast of developments that can affect their work. Consequently, these functions have been transferred largely to a specialist who can not only gather information efficiently but also manage information to serve the needs of the users. This specialist must be technically competent and have the ability to evaluate information.

Although the benefits of on-line retrieval are clearly evident, there is concern that users should be in direct contact with the information they need. The phrases friendly systems, friendly interfaces, and user-oriented systems are beginning to be heard. At one time, users went directly to their information sources with little effort. In fact, many chemists had their own subscriptions to *Chemical Abstracts* and other secondary materials. They considered it part of their obligation to maintain knowledge of developments relative to their work. As the generation of published literature escalated and the journals proliferated, the sheer volume of material appeared to be endless. Chemists found their task impossible using the old method of manual scanning. The advent of computerized current-awareness tools gave them a way to review literature and indeed made it possible once again to keep abreast of new developments. However, information retrieval continues to be a difficult task for average users without the assistance of information intermediaries. Work is in progress to develop systems better oriented to users.

Before 1950, graduating chemists and engineers could look forward to applying their training throughout their careers. Now there is the potential of technical obsolescence of scientists and engineers. To thwart this possibility, the aggressive use of information is essential.

The following section addresses the matter of scientific and technical information retrieval using traditional as well as computerized methods and resources.

Bibliography

J. A. Luedke and co-workers, "Numerical Databases and Systems" in M. E. Williams ed., *Annu. Rev. Inf. Sci. Technol.* **12,** 119–181 (1977).

J. M. Morris and E. A. Elkins, *Library Searching: Resources and Strategies with Examples from the Environmental Sciences,* Jeffrey Norton Publishers, Inc., New York, 1978.

B. H. Weil, "Authorized Services for Supplying Photocopies and for Collection of Payments for In-House Photocopying under the New Copyright Law," in N. B. Glick and F. Simora, eds., *The Bowker Annual of Library and Book Trade Information,* 23rd ed., R. R. Bowker Company, New York, 1978, pp. 33–41.

M. E. Williams, "On-line Retrieval—Today and Tomorrow," *On-Line Rev.* **2**(4), 353–366 (1978).

SCIENTIFIC AND TECHNICAL INFORMATION FOR BASIC AND APPLIED RESEARCH

Studies of scientific and technical information users disclose interesting findings. Radwin found that only 10.7% of library users needed exhaustive information; 39.9% needed single facts or data; and 44.3% needed background information. Direct use of the literature produced maximum idea generation. However, problem solvers found three information-gathering techniques to be equivalent: personal interaction, literature use, and experimentation and analysis. This duality of information needs for idea generation and for problem solving is similar to Garfield's description of scientific and technical information needs as recovery and discovery, or Radwin's description of known and unknown needs.

Two-part division is used here, dealing first with the fact-finding or problem-solving needs of chemists at bench and production activities, and then with the comprehensive and conceptual needs for basic research or research planning.

Selected Bibliography of User Studies

T. J. Allen, *Managing the Flow of Technology: Technology Transfer and the Dissemination of Technological Information Within the R&D Organization,* MIT Press, Cambridge, Mass., 1977.

A. K. Chakrabarti, "Information Use and Training of Industrial Scientists and Engineers," *Libr. Sci. Slant Doc.* **15,** Paper N (June, 1978).

S. Crawford, "Information Needs and Uses" in M. E. Williams, ed., *Annu. Rev. Inf. Sci. Technol.* **13,** 61–81 (1978).

E. Garfield, C. E. Granito, and A. E. Petrarca, "Information Retrieval Services and Methods" in A. Standen, ed., *Kirk-Othmer Encyclopedia of Chemical Technology,* 2nd ed., Supplement Volume, Interscience Publishers, a division of John Wiley & Sons, Inc., New York, 1971, pp. 510–535.

K. Holt, "Information and Need Analysis and Idea Generation," *Res. Manage.* **18**(3), 24–27 (1975).

J. Martyn, "Information Needs and Uses" in C. A. Cuadra ed., *Annu. Rev. Inf. Sci. Technol.* **9,** 3–23 (1974).

P. S. Nagpaul and S. Pruthi, "Problem-Solving and Idea-Generation in R&D: the Role of Informal Communication," *R&D Manage.* **9**(3), 147–149 (1979).

E. Radwin, "Field Survey of Information Needs of Industry Sci/Tech Users" in C. W. Husbands and R. Tighe, eds., *Information Revolution: Proceedings of the 38th Annual Meeting of the American Society for Information Science,* American Society for Information Science, Washington, D.C., **12,** 41–42 (1975).

B. H. Weil, "Benefits from Researcher Use of the Published Literature at the Exxon Research Center" in E. Jackson, ed., *Special Librarianship; A New Reader,* Scarecrow Press, Metuchen, N.J., 1980.

F. W. Wolek, "Uses and Benefits of Technical Information Systems," *Res. Manage.* **20**(5), 37–41 (1977).

Bibliography of Chemical Literature Sources

Some readers may wish to learn more about chemical information resources. The following bibliography lists some current literature guides:

R. T. Bottle, *Use of the Chemical Literature,* 3rd ed., Butterworths Publishers, Inc., Woburn, Mass., 1979.

C. Chen, *Scientific and Technical Information Sources,* MIT Press, Cambridge, Mass., 1977; an extensive bibliography with citations of book reviews and reference librarians.

R. E. Maizell, *How to Find Chemical Information: A Guide for Practicing Chemists, Teachers, and Students,* Wiley-Interscience, New York, 1979; descriptive guide for chemists, mixed with sections for librarians.

D. B. Owen and M. M. Hanchey, *Indexes and Abstracts in Science and Technology: A Descriptive Guide,* Scarecrow Press, Metuchen, N.J., 1974.

T. P. Peck, ed., *Chemical Industries Information Sources (Management Information Guide Series 29),* Gale Research Company, Detroit, Mich., 1978; lists resources in chemistry, chemical engineering, and associated industries; this is a wide-ranging guide to organizations as well as literature, including classic literature.

S. H. Wilen, *Use of the Chemical Literature, An Introduction to Chemical Information Retrieval,* ACS Audio Course, American Chemical Society, Washington, D.C., 1978.

H. M. Woodburn, *Using the Chemical Literature: A Practical Guide (Library and Information Science Series,* Vol. 11), Marcel Dekker, Inc., New York, 1974; a self-teaching tool; this informative guide to some main resources for obtaining chemical information discusses manual retrieval and current-awareness techniques in a clear and practical manner; it deals only with those resources known to cause problems to users.

Information in the Laboratory

At the bench level, chemists are looking for facts, often urgently because an experiment is already in progress. Besides facts, experimental procedures or techniques are used at the bench. Until now the information resources for these needs were in traditional book form and were usually kept in the chemist's personal library or, in the case of larger data compilations, series on techniques, and encyclopedias, in the organization's library.

There is now a trend toward delivering factual, handbook information in electronic form via the laboratory's computer terminal, already in use for automated instrument control and data analysis. For the chemical industry, delivery of handbook information is workable if the data, equations, and the like are accompanied by design modules that allow retrieved information to be manipulated to produce an answer (2). In contrast, simple electronic reference systems are not as useful.

The discussion here concentrates on a selected list of universally available, printed resources. The literature guides listed in the bibliography and the reader's reference librarian or information specialist will be helpful in locating the most appropriate and up-to-date resources for the reader's particular needs.

Handbooks. Handbooks are the most convenient reference tools for the laboratory. They contain just the kind of information required in the middle of things—physical constants and mathematical tables. The data may or may not be evaluated but they are usually reliable—the hard core of chemistry.

Bibliography

American Chemical Society, Committee on Analytical Reagents, *Reagent Chemicals: American Chemical Society Specifications,* 5th ed., American Chemical Society, Washington, D.C., 1974.

J. A. Dean, ed., *Lange's Handbook of Chemistry,* 12th ed., McGraw-Hill Book Co., New York, 1979; includes physical constants, thermodynamic properties, and mathematical tables and equations; some self-instructional units; designed for desk use.

A. J. Gordon and R. A. Ford, *The Chemist's Companion; A Handbook of Practical Data, Techniques, and References,* Wiley-Interscience, New York, 1973; designed for easy use in the lab.

F. A. Lowenheim and M. K. Moran, *Faith, Keyes, and Clark's Industrial Chemicals,* 4th ed., John Wiley & Sons, Inc., New York, 1975.

R. H. Perry and C. H. Chilton, *Chemical Engineers' Handbook,* 5th ed., McGraw-Hill Book Co., New York, 1973; completely revised to assist engineers in producing efficient designs based on scientific advances.

J. Pinkava, *Handbook of Laboratory Unit Operations for Chemists and Chemical Engineers,* Gordon and Breach Science Publishers, New York, 1971.

G. J. Shugar, *Chemical Technicians' Ready Reference Handbook,* McGraw-Hill Book Co., New York, 1973.

R. C. Weast, ed., *CRC Handbook of Chemistry and Physics,* 59th ed., CRC Press, Inc., West Palm Beach, Florida, 1978–1979; original edition 1918; lists sources of critical data; revised annually; physical constants, spectral data, thermodynamic properties, conversion factors, references to numeric data projects, useful index.

There are many highly specialized handbooks. The *Subject Guide to Books in Print* (R. R. Bowker, New York, annual publication) and *Handbooks and Tables in Science and Technology* (R. H. Powell, ed., Oryx Press, 1980) are useful in identifying relevant ones in individual areas. A few examples are the following:

G. L. Baughman, *Synthetic Fuels Data Handbook,* 2nd ed., Cameron Engineers, Inc., Denver, Colorado, 1978; synthetic fuels from oil shale, coal, and oil sands.

C. A. Harper, *Handbook of Plastics and Elastomers,* McGraw-Hill Book Co., New York, 1975.

I. Mellan, *Industrial Solvents Handbook,* Noyes Data Corp., Park Ridge, N.J., 1977; useful in solvent selection or replacement.

Synthetic Organic Chemical Manufacturers Association, *SOCMA Handbook; Commercial Organic Chemical Names,* American Chemical Society, Washington, D.C., 1965; lists 6300 industrial organic compounds, giving structure, nomenclature, and Chemical Abstracts Service Registry Number.

Physical Data Compilations. Using data compilations can save chemists' time if they provide quality information (3). The institutions that produce these collections spend considerable effort in evaluating data. The complex organizational schemes designed to accommodate updates require the user to read the introduction carefully.

The following list represents only a small sample of the useful physical data collections; these have proved their value over many years.

International Compendium of Numerical Data Projects, a Survey and Analysis, CODATA-The Committee on Data for Science and Technology of the International Council of Scientific Unions, Springer-Verlag, New York, 1969.

E. W. Washburn, ed., *International Critical Tables of Numerical Data—Physics, Chemistry and Technology,* published for the National Research Council by McGraw-Hill Book Co., New York, 1926–1933; 7 data volumes and 1 index volume; an old but significant and trusted data source.

Landolt-Bornstein Zahlenwerte und Funktionen aus Physik, Chemie, Astronomie, Geophysik und Technik, 6 Aufl., Springer-Verlag, Berlin, 1950–.

K. H. Hellwege, ed., *Landolt-Bornstein Zahlenwerte und Funktionen aus Naturwissenschaften und Technik,* Neue Serie, Springer-Verlag, New York, 1961–; this series contains reliable data but no index: to locate information one selects a volume by reading the name on the spine and identifies the likely table from the contents; in the new series, side-by-side English and German are used; the old series is entirely in German.

The National Standard Reference Data System (NSRDS) of the National Bureau of Standards (NBS) was established to coordinate the data-compiling activities of several governmental agencies. The publications from this program are listed in the *Handbook of Chemistry and Physics,* and much of the information is now published in the *Journal of Physical and Chemical Reference Data,* as well as in NSRDS-NBS publications.

The Center for Information and Numerical Data Analysis and Synthesis (CINDAS), Purdue University, Y. S. Touloukian, director, operates the Thermophysical Properties Research Center.

Thermophysical Properties of Matter, 13 vols., IFI/Plenum, New York, 1970–1977; covers thermal conductivity, specific heat, radiative properties, thermal diffusivity, viscosity, and thermal expansion; supplements are being published.
Thermophysical Properties Research Literature Retrieval Guide, Y. S. Touloukian, IFI/Plenum, New York, 1967: *Basic Ed.,* 3 volumes, 1967; *Supplement I,* 6 volumes, 1973; *Supplement II,* 6 volumes, 1979; the *Retrieval Guide,* which is complementary to the TRC data tables, provides extensively indexed access to world literature on thermophysical properties; literature from 1920 to mid 1964 is covered in the *Basic Ed.;* mid 1964 to 1971 in *Supplement I,* and 1971–1977 in *Supplement II.*

Thermodynamics Research Center (TRC) at Texas A&M University, B. J. Zwolinski, director, participates in NSRDS and has two major projects. The arrangement of these publications is complex.

B. J. Zwolinski and co-eds., *Comprehensive Index of API 44—TRC Selected Data on Thermodynamics and Spectroscopy, TRC Publication #100,* 2nd ed., 1974; conventional index to all TRC publications.
Selected Values of Properties of Hydrocarbons and Related Compounds; Thermodynamics Research Center, American Petroleum Institute Research Project 44; this series provides evaluated physical, thermodynamic, and spectral data for thousands of compounds.
Selected Values of Properties of Chemical Compounds; Thermodynamics Research Center Project; covers organic compounds other than the hydrocarbons included in *API 44.*

Bibliography

JANAF (Joint Army–Navy–Air Force) Thermochemical Tables, 2nd ed., NSRDS-NBS 37, Office of Standard Reference Data, National Bureau of Standards, Superintendent of Documents, U.S. Government Printing Office, Washington, D.C., 1971; critical compilation of tables of thermodynamic properties of propellant–combustion products and inorganics; supplements published in *J. Phys. Chem. Ref. Data.*
Sadtler Standard Spectra (INFRARED), Sadtler Research Laboratory, Philadelphia, Pa.; regularly updated looseleaf service.
A. Seidell, *Solubilities of Inorganic and Metal-Organic Compounds,* 4th ed., Vol. 1, D. Van Nostrand Co., Inc., New York, 1958; Vol. 2, American Chemical Society, Washington, D.C., 1965; unevaluated; aqueous and nonaqueous solvent systems; references to data sources.
H. Stephen and T. Stephen, *Solubilities of Inorganic and Organic Compounds,* 2 vols., Pergamon Press, Ltd., Oxford, Eng.; distributed by the Macmillan Co., New York, 1963.
G. W. C. Kaye and T. H. Laby, *Tables of Physical and Chemical Constants and Some Mathematical Functions,* 14th ed., Longman Group Ltd., London, Eng., 1973; all tabulated values now in SI units; aim is to provide data in a convenient size and at a moderate price.

Encyclopedias and Treatises.   Encyclopedias provide background information. The treatises, by dealing in-depth, can help the nonspecialist become more knowledgeable. Treatises generally are descriptive, contain evaluated data, discuss methodologies, and are likely to have better indexes than handbooks or monographs.

C. H. Bamford and C. F. H. Tipper, *Comprehensive Chemical Kinetics,* 18 vols., American Elsevier Publishing Company, New York, 1969–1977; multivolume sections; coverage is thorough; Section 1—Practice and Theory; Section 2—Homogeneous Decomposition and Isomerization Reactions; Section 3—Inorganic Reactions; Section 4—Organic Reactions; Section 5—Polymerization Reactions; Section 6—Oxidation and Combustion Reactions; Section 7—Selected Elementary Reactions.
D. H. R. Barton and W. D. Ollis, eds., *Comprehensive Organic Chemistry: The Synthesis and Reactions of Organic Compounds,* 6 vols., Pergamon Press, New York, 1979; the expanding literature created a need for a new, rapidly-published work in organic chemistry. One technique the editors used to accomplish this goal was not to treat theoretical organic chemistry separately, since this area changes first. This is a six-

volume work, with the sixth containing the indexes, with references to review articles. The indexes are organized by formula, subject, author, reaction, and reagents, and provides pointers to the contents of volumes 1–5, the references in those volumes, and significant literature through mid 1978.

Chemical Technology: An Encyclopedic Treatment; The Economic Application of Modern Technological Developments, 8 vols., Barnes & Noble, Inc., New York, 1968–1973; for lay persons and technologists.

S. Coffey, ed., *Rodd's Chemistry of Carbon Compounds, A Modern Comprehensive Treatise,* 2nd ed., Elsevier Publishing Company, New York, 1964–; five multipart volumes arranged using the Richter classification.

Dictionary of Organic Compounds (Heilbron): The constitution and physical, chemical, and other properties of the principal carbon compounds and their derivatives together with relevant literature references, 4th ed., 5 vols., Oxford University Press, New York, 1965; updated by supplements; constitution and physical and chemical properties as extracted from the literature.

J. E. Faraday, ed., *Encyclopedia of Hydrocarbon Compounds,* 13 vols., Chemical Publishing Co., New York, 1946–1954.

Kirk-Othmer Encyclopedia of Chemical Technology, 2nd ed., A. Standen, ed., 1963–1972; 3rd ed., 1978–, M. Grayson and D. Eckroth, eds., Wiley-Interscience, New York; 12 volumes published by end 1980; annual indexes; 25 vols to be published in 3rd ed., by 1984 including cumulative index.

I. M. Kolthoff and P. J. Elving, eds., *Treatise on Analytical Chemistry,* John Wiley & Sons, Inc., New York, 1959–; a comprehensive review series in three parts: I—Theory and Practice; II—Analytical Chemistry of Inorganic and Organic Compounds; III—Analytical Chemistry in Industry; started in the 1960s and continuing.

Encyclopedia of Chemical Processing and Design, Marcel Dekker, Inc., New York, 1976 (J. J. McKetta and W. A. Cunningham, eds.); intended to assist designers and developers of chemical processes and products in practical design problems; 11 volumes published by end of 1980.

J. W. Mellor, *A Comprehensive Treatise on Inorganic and Theoretical Chemistry,* Longman's, Green & Co., Ltd.; John Wiley & Sons, Inc., New York, 16 vols. (main series), 1922–1937; supplements, 1956–1972; arranged by periodic table, this work presents information in the order: history, preparation, properties, hydride, oxide, halides, sulfide, sulfate, carbonate, nitrate, phosphate, references; each primary and supplementary volume has an index; volume 16 is a general index to the main series.

E. Josephy and F. Radt, eds., *Elsevier's Encyclopedia of Organic Chemistry,* Springer-Verlag, Berlin, 1948–1965; based on structural skeleton, this encyclopedia offers compact presentation of data; updated by supplements.

F. D. Snell and L. S. Ettre, *Encyclopedia of Industrial Chemical Analysis,* 20 vols., Wiley-Interscience, New York, 1966–1974; Vols. 1–3 cover general techniques, Vols. 4–19 are specific, and Vol. 20 is the index to Vols. 4–19; coverage includes individual compounds of commercial significance (acetone, dioxane), elements and their compounds (aluminum), chemical classes (amines), compounds by end use (adhesives), and compounds by industrial use (food products).

J. C. Bailar, Jr., H. J. Emeleus, R. Nyholm, and A. F. Trotman-Dickenson, eds., *Comprehensive Inorganic Chemistry,* 5 vols., Pergamon Press, New York, 1973; the presentation in each compound or family is compact with much data in tabular form and references as footnotes; the index is quite complete.

Ullmann's Encyklopaedie der Technischen Chemie, 4th ed., Verlag Chemie International, Deerfield Beach, Fl., 1972–; 25 vols. to be published in the 4th ed.

Beilstein and Gmelin.   The Beilstein and Gmelin chemical treatises hold a special place in the minds of chemists because they present evaluated information on a comprehensive scale. Both are compound-based but have different organizational schemes. The value of these handbooks is immense, and retrieval is rapid; instructional information is available for each.

Beilstein's Handbuch der Organischen Chemie, simply known as *Beilstein,* is a comprehensive reference source of carbon-containing compounds. First published in 1881–1882 in two volumes by F. K. Beilstein, the handbook now comprises more than 190 volumes. More than 100 chemists and physicists are employed at the Beilstein Institute to review critically all primary publications containing information on carbon compounds. The editor is R. Luckenbach. In addition, the chemical literature is evaluated, and data are checked to ensure that misleading or erroneous results are not documented. *Beilstein* is a key source to factual chemical information, free from the inaccuracies and redundancies of the open chemical literature.

The *Beilstein Handbook* began as 27 volumes of the *Basic Series,* using the Beilstein system to order carbon compounds logically. This system was developed by P. Jacobson and B. Prager in 1907 and continues to be used as the organic chemical classification system by which information is collected and presented. Details of the Beilstein system are found in the opening pages of the *Basic Series'* Volume 1, which should be consulted to ensure proper usage of the handbook. Four series of supplemental volumes have been added to the *Basic Series* since 1910, and a fifth series covering the literature from 1960 to 1979 is currently being prepared. Continuity is one of the most valued features of the Beilstein system. New information on a compound already reported is published in the same volume of later *Supplement Series.* Thus, once the volume and system number are identified, all other information in *Beilstein* on that compound can be located easily.

Beilstein gives specific data on and references to nearly four million (4×10^6) compounds of known constitution. The description of such compounds includes their composition, configuration, natural occurrence and isolation from natural products, preparation and purification, structural and energy parameters, physical properties, chemical properties, their characterization and analysis, and their salts and addition compounds. A booklet entitled *How to Use Beilstein* is available free of charge from the publisher, Springer-Verlag, Berlin, Heidelberg, and New York. Woodburn (4) includes a helpful chapter on using *Beilstein.*

The *Gmelin Handbuch der Anorganischen Chemie*, or simply *Gmelin,* is the most authoritative and comprehensive treatise on inorganic compounds. The first edition was published in 1817 by L. Gmelin. The current, eighth edition of 365 volumes is published in Frankfurt by the Gmelin-Institut für Anorganische Chemie und Grenzgebiete, an institute of the Max Planck Society for the Advancement of Science. The editor is E. Fluck. Like *Beilstein, Gmelin* is a compilation of data from primary literature that has been critically evaluated and presented in coherent groups of related subject matter.

Each volume and its supplement discuss an element and its compounds, their formation and occurrence, methods of preparation, and physical and chemical properties. The classification system used in *Gmelin* is based on a numerical sequence of the elements by system number. Information about a compound is found under that element in the compound having the highest system number. One key to all the elements and compounds of the eighth edition is the formula index, now in preparation. The Gmelin Institute is progressively translating its entire collection from German to English; the complete handbook eventually will be available in English.

The value of the *Gmelin Handbook* must not be underestimated. The current edition contains reevaluated material from the older editions and thus represents a comprehensive documentation of the entire body of inorganic chemical literature. Material is presented in a form that allows comparative assessments and includes references to allied concepts and critical reviews. The handbook is described in detail in the Institute's brochure *Was ist der Gmelin?,* available in either English or German from the Gmelin Institute, Frankfurt. H. M. Woodburn has described the original scope and use of the handbook (5).

Bibliography

F. Beilstein, *Handbuch der Organischen Chemie,* 4th ed., Springer-Verlag, Berlin.
How to use Beilstein, Beilstein Handbook of Organic Chemistry, Beilstein Institute, Springer-Verlag, New York.
R. Luckenbach, "Der Beilstein," *Chemtech* **9**(10), 612–621 (1979).
H. M. Woodburn, *Using the Chemical Literature: A Practical Guide* (*Library and Information Science Series,* Volume 11), Marcel Dekker, Inc., New York, 1974.
Gmelin's Handbuch der Anorganischen Chemie, 8th ed., Springer-Verlag, New York.

Experimental Methods.   Chemists have a long tradition of publishing methodological compilations. These make it easy to locate synthetic preparations or discussions of laboratory techniques at a more practical level than found in encyclopedias.

Weissberger's *Techniques of Chemistry* series (6), a successor to his *Technique of Organic Chemistry,* is organized into several multipart volumes: Vol. 1—Physical Methods of Chemistry, A. Weissberger and B. W. Rossiter, eds., 1971–1977; Vol. 2—Organic Solvents: Physical Properties and Methods Purification, J. A. Riddick and W. B. Bunger, eds., 1971; Vol. 3—Photochromism, G. H. Brown, ed., 1971; Vol. 4—Elucidation of Organic Structures by Physical and Chemical Methods, K. W. Bentley and G. W. Kirby, eds., 1972–1973; Vol. 5—Technique of Electroorganic Synthesis, N. L. Weinberg, ed., 1974–1975; Vol. 6—Investigation of Rates and Mechanisms of Reactions, E. S. Lewis and G. G. Hammes, eds., 1974; Vol. 7—Membranes in Separation, S. Hwang and K. Kammermayer, eds., 1975; Vol. 8—Solutions and Solubilities, M. R. J. Dack, ed., 1975–1976; Vol. 9—Chemical Experimentation under Extreme Conditions, B. W. Rossiter, ed., 1980; Vol. 10—Applications of Biochemical Systems in Organic Chemistry, J. B. Jones, C. J. Sih, and D. Perlman, eds., 1976; Vol. 11—Contemporary Liquid Chromatography, R. P. W. Scott, ed., 1976; Vol. 12—Separation and Purification, E. S. Perry and A. Weissberger, eds., 1978; Vol. 13—Laboratory Engineering and Manipulation, A. Weissberger and E. S. Perry, eds., 1979; Vol. 14—Thin Layer Chromatography, J. G. Kirchner, ed., 1978; Vol. 15—Theory and Applications of Electron Spin Resonance, W. Gordy, ed., 1979.

E. Muller, ed., *Houben-Weyl Methoden der Organischen Chemie,* 4th ed., Thieme, Stuttgart, FRG, 1952–; originated early in the century as a critical survey of laboratory methods; gives physical and analytical methods as well as preparations, transformations, and theory.

Organic Reactions, John Wiley & Sons, Inc., New York, 1942–; annual series providing critical discussion of widely used organic reactions.

Organic Syntheses, John Wiley & Sons, Inc., New York, 1921–; annual volumes published since 1921; the ten-year collective volumes are revised as required; emphasis is now on model procedures for reaction types, rather than on preparations of specific organic compounds.

Theilheimer's *Synthetic Methods of Organic Chemistry,* (S. Karger), a series based on reaction schemes, and Buehler and Pearson's *Survey of Organic Synthesis* (Wiley) are series that help the organic chemist locate experimental methodology rapidly.

Dictionaries.   There are many technical dictionaries used to define specific terms and, to a limited degree, chemical structures. In chemistry, the line between dictionaries and encyclopedias is thin but the term dictionary here refers to the one-volume works. Technical foreign language dictionaries are not covered here.

Bibliography

H. Bennett, ed., *Concise Chemical and Technical Dictionary,* 3rd ed., Chemical Publishing Company, New York, 1974.
T. C. Collocott and A. B. Dobson, eds., *Chambers Dictionary of Science and Technology Terms,* successor to *Chambers Technical Dictionary,* W&R Chambers, Ltd., Edinburgh, UK, 1974.
G. L. Clark and G. G. Hawley, eds., *The Encyclopedia of Chemistry,* 2nd ed., Reinhold Publishing Corp., New York, 1966.
D. M. Considine, ed., *Van Nostrand's Scientific Encyclopedia,* 5th ed., Van Nostrand Reinhold Company, New York, 1976; in addition to chemistry, this covers physics, materials sciences, energy technology, earth and space sciences, life science, mathematics, and information science; references included.
W. Gardner, E. I. Cooke, and R. W. I. Cooke, eds., *Handbook of Chemical Synonyms and Trade Names,* 8th ed., CRC Press, Cleveland, Ohio, 1978.

J. Grant, ed., *Hackh's Chemical Dictionary,* 4th ed., McGraw-Hill Book Co., New York, 1969.

D. N. Lapedes, ed., *McGraw-Hill Dictionary of Scientific and Technical Terms,* 2nd ed., McGraw-Hill Book Co., New York, 1978.

G. G. Hawley, ed., *Condensed Chemical Dictionary,* 9th ed., Van Nostrand-Reinhold Corp., New York, 1977.

M. Windholz, S. Budavari, L. Y. Stroumtsos, and M. Noether Fertig, eds., *The Merck Index,* 9th ed., Merck & Co., Inc., Rahway, N.J., 1976; descriptive information on 10,000 chemicals, drugs, and biologicals.

Reviews. One rapid way for chemists to keep current in several areas or to bring their knowledge to the state-of-art level is to read reviews. In fact, on-line data bases can be searched using REVIEW as a keyword or document type. Publishers even provide indexes to reviews. The Institute for Scientific Information publishes one of these semiannually and cumulates it annually in *Index to Scientific Reviews.*

There are several review journals: *Accounts of Chemical Research,* American Chemical Society; *Angewandte Chemie—International Edition,* in English, Verlag Chemie; *Chemical Society Reviews,* The Chemical Society; *Chemical Reviews,* American Chemical Society; and *CRC Critical Review Series,* CRC Press.

A number of review series are published, often entitled *Advances in . . . ,* or *Progress in . . . .* The Chemical Society publishes two such series, *Annual Reports on the Progress of Chemistry,* and *Specialist Periodical Reports.* The Society of the Chemical Industry publishes a similar series called *Report on the Progress of Applied Chemistry.*

Miscellaneous. There are more sources of information than we can discuss in one article. Some of these are governmental reports, dissertations, monographs, standards, technical brochures, catalogues, and conference papers. Because abstracting and indexing tools provide access to these also, it is now no more difficult to identify a conference paper than to identify an article in the *Journal of the American Chemical Society.*

Technical Information, Retrospective and Comprehensive

So far this article has discussed techniques of information retrieval, the importance of information in all aspects of the chemist's work, and the kinds of resources used for fact finding. Reviewed here are some main resources for retrospective searching—exhaustive and background, manual and computerized—and current awareness as printed bulletins or personalized, computer-produced, SDI profiles. The review begins with a discussion of search services.

Lockheed–DIALOG. Lockheed Information Service (LIS) in Palo Alto, California, was established in 1965 as a division of Lockheed Missiles and Space Corporation. LIS designed a retrieval system called RECON (REmote CONsole) for the document-citation collection of the National Aeronautics and Space Administration (NASA) and further developed it into the DIALOG on-line information system. Commercial operation began in 1972 providing access to three data bases—ERIC (Educational Resources Information Center), CAIN (Cataloging and Indexing data base of the National Agricultural Library), and PANDEX (the on-line version of the *PANDEX Current Index to Scientific and Technical Literature*). The current collection contains more than 100 data bases in a wide range of disciplines, including directories and statistical files as well as bibliographic data bases, and with access to over thirty million records in total. The cost of searching is based on connect time to the particular data base, that is, the actual time elapsed from sign-on to sign-off from a certain computer file. No charge is made until the system is used.

System Development Corporation–ORBIT. System Development Corporation (SDC), founded in 1956 in Santa Monica, California, designed a system called COLEX for the Foreign Technology Division of the United States Air Force System Command. This system developed into a series of ELHILL programs and finally into ORBIT (On-line Retrieval of Bibliographic Information Timesharing). Commercial operation of the SDC Search Service began in 1973 offering access to the three data bases, ERIC, CHEMCON (Chemical Condensates), and MEDLINE (Medlars on line). Currently there are 65 data bases providing access to over twenty-five million records. On-line searching is charged by connect hour.

Bibliographic Retrieval Service–STAIRS. Established in 1976 in Scotia, N.Y., the Bibliographic Retrieval Service (BRS) is the newest of the United States on-line search services. It was formed to offer on-line service affordable to small organizations and public organizations such as state libraries. BRS uses a modification of the STAIRS (Storage and Information Retrieval System) software system, developed by International Business Machines (IBM). It provides access to 30 data bases, 9 of which are unique to BRS, bringing the total number of searchable records to nearly sixteen million. Coverage is expanding to include on-line access to standard reference works. BRS files are used on a subscription basis, ie, the user is entitled to a specific number of connect hours that may be used at any time during the year. Reduced rates are offered for increased usage. These rates, however, are exclusive of communications charges, any applicable data-base royalty charges, and optional services such as off-line searching, printing, and SDI.

National Library of Medicine–MEDLARS. The National Library of Medicine in Bethesda, Maryland, began experiments with on-line bibliographic searching in 1967. Time-consuming, expensive batch searches of their *Index Medicus* data base were run using the Medical Literature Analysis and Retrieval System (MEDLARS), a system devised originally to produce *Index Medicus.* Because MEDLARS continued to grow with the expansion of the biomedical literature and the increased number of demand searches, a faster and inexpensive system was needed. SDC was hired to develop the on-line searching portion of MEDLARS; the ELHILL III system has emerged as the operating system since 1975. Currently there are 18 data bases accessible on line, 14 of which are available only through NLM. Five data bases available for off-line searching contain citations, and often abstracts, of the biomedical literature going back to the mid to late 1960s. A total of five million records are accessible. NLM is relatively inexpensive and offers reduced rates for usage outside of prime time.

Department of Energy–RECON. The DOE–RECON (Department of Energy–REmote CONsole) system was initially developed by Lockheed for NASA. The Atomic Energy Commission purchased the system to support the *Nuclear Science Abstracts* data base, and the system was installed at Oak Ridge National Laboratory (ORNL) in 1970–1971, where it operated experimentally for about four years while the on-line retrieval software was improved. Dial-up access to DOE–RECON has been available since 1975. The current version contains 21 unclassified data bases, including a few trial data bases that will be retained if use is sufficient. Most of the data bases are prepared by governmental offices and are unique to DOE–RECON. The DOE–RECON system differs from the commercial search services in that it is restricted to use by DOE personnel, holders of DOE-sponsored research contracts, and other federal agencies engaged in work related to DOE's programs.

Canada Institute for Scientific Information–CAN–OLE.   The Canada Institute for Scientific and Technical Information (CISTI) was formed by the National Research Council of Canada in 1974 to provide Canadian researchers, technologists, and industries with scientific and technical information. The Institute operates a range of information services, including a computerized current-awareness service (CAN–SDI) and an on-line reference system (CAN–OLE). In cooperation with the U.S. National Library of Medicine, it provides on-line access in Canada to the medical literature (MEDLINE). CAN–OLE currently contains 12 data bases, which can be interactively searched in either English or French. Paper copies of the references retrieved may be ordered from CISTI. Billing for the service is on a connect-hour basis. Through the DATAPAC communications network of the Trans-Canada Telephone System, there is access to these information services from over 50 cities.

QL Systems Ltd.–Shared Information Service.   QL Systems Ltd. was an outgrowth of the QUIC–LAW Project at Queen's University, Ontario, Canada. The project was established in 1968 to investigate possible use of computers by lawyers, particularly in legal research. Owing to its rapid growth in the early seventies, the project was incorporated as a separate entity outside the University. Since 1973, QLS has been dedicated to developing and maintaining the first Canadian commercial information-retrieval system, the Shared Information Service. Over 30 data bases are currently available, including some full-text files. The subject coverage ranges from the environment to business to law with most of the legal and governmental files in both English and French. On-line searching is billed per search, for each data-base selection, and by a communications charge per on-line hour in the United States; some data bases are also subject to a royalty charge per on-line hour.

EURONET–DIANE.   EURONET is the data-transmission network set up by the postal and telecommunications authorities of European Economic Communities (EEC) to provide on-line access to host computers and their data bases. DIANE (Direct Information Access Network for Europe) is a group of host information services that offer their data bases through EURONET. A few of these are BLAISE (British Library Automated Information Service); Datacentralen; DIMDI (Deutches Institut für Medizinische Dokumentation und Information); EPO; Infoline; and IRS (Information Retrieval Service), the oldest European service.

Chemical Abstracts Service.   One might think it unnecessary to discuss *Chemical Abstracts (CA)* in an article on information retrieval for chemists. It is the one tool chemists learn about as undergraduates. However, beginning in the early 1960s Chemical Abstracts Service (CAS) formulated new methods for carrying out its mission—the abstracting and indexing of the world's chemical literature. The volume of published literature was exploding, and the power of the computer was becoming an everyday reality.

The full effect of the growth in scientific and technical literature is clear when we see that in 1979 Chemical Abstracts Service published nearly 437,000 abstracts and covered 14,000 periodicals. This means that approximately 16,800 abstracts were issued every two weeks; no one chemist can cope with that. Even the indexes have grown in depth and size. The *8th Collective Index* comprises 35 books; the *9th Collective Index,* 57 books; and the *10th Collective Index* (1977–1981) is expected to have between 71 and 75 volumes of indexes alone.

International cooperation is the first of the three principles CAS adopted to manage information in a changing environment. The chemical societies of The Federal Republic of Germany, the United Kingdom, France, and Japan provide abstracts from their national chemical publications or pay a share of production costs, in exchange for certain use and distribution rights. The second principle is the data-base concept, through which CAS has established multiple, compatible files that allow new and timely services. CAS maintains three types of files: bibliographic information; abstract text; and data elements for subject indexes. The third principle is to use unique, noninformative, identifying numbers for chemicals for *CA* production (structure, nomenclature), and for information retrieving and linking—Registry Numbers.

CAS Indexes.   *Chemical Abstracts* has indexes of varying depth and type. In 1963 *CA* added a permuted keyword–subject index and patent concordance to its author and numerical patent indexes for the weekly issues.

The volume semiannual indexes and the collective indexes, now issued every five years, are much more extensive. These indexes are as follows: Author Index, Formula Index, Index of Ring Systems, Numerical Patent Index, Patent Concordance, General Subject Index, and Chemical Substance Index.

It is important to consult the *CA Index Guide,* first issued in 1969, because cross references, scope notes, and synonyms are no longer printed in the subject or substance indexes. The *Registry Handbook* and its *Update,* arranged in Registry Number order, tell the *CA* systematic name, the molecular formula, and any changes in registration or cross references. Occasionally, multiple registrations need to be resolved, or changes made based on further structural elucidation. The *Parent Compound Handbook* replaces the old *Ring Index.* It has a broader scope and six routes of access: the *Ring Analysis Index,* the *Ring Substructure Index,* the *Parent Name Index,* the *Wiswesser Line Notation Index,* the *Parent Formula Index,* and the *Parent Registry Number Index.*

CAS Computer Searching.   *CA* indexes can also be searched by on-line, interactive, computer techniques from several commercial and national services (SDC, Lockheed, BRS in the U.S.; EURONET–DIANE in Europe; and CISTI in Canada), all based on CA SEARCH. Although the time span accessible (late 1960s or early 1970s to the present) is limited compared to the manual search of indexes back to 1907, computers can search rapidly, allow for iterative and heuristic querying, search across information types (author and subject and patent number) to search information types not accessible in the manual indexes (organizational author, document type, word fragments), and provide substructure searching based on nomenclature and registry information. In addition, the use of Boolean logic enables one to structure complex queries, in contrast to searching printed indexes one heading at a time. Another limitation of computer-produced search results is the lack of abstracts.

CAS Current Awareness.   CAS has a number of current-awareness services. When individual issues became unwieldy, *CA* established five section groups, which could be subscribed to separately. These divide the 80 sections into biochemistry, organic chemistry, macromolecular chemistry, applied chemistry and chemical engineering, and physical and analytical chemistry.

Chemical Titles (*CT*), the first computer-produced publication, is a general publication that has a permuted key-word-in-context (KWIC) Index to article titles from over 700 journals of pure and applied chemistry and chemical engineering. It also contains a bibliographic section organized by journal and an author index. *CT* is a timely publication since CAS prepares it from galley proofs.

Personalized SDI profiles based on *CA,* a computer technique discussed earlier, are available in many organizations, internal, government-sponsored, and commercial.

CA Selects is a new service that reflects CAS's responsiveness to consumer demands and the flexibility of its automated systems. *CA Selects* are topical current-awareness bulletins that draw across *CA* sections, include the abstracts, and are printed with *CA* quality. The bulletins average 136 abstracts on 12 pages. The computer-searching profiles use all the information in the CAS data base: bibliographic data, keyword phrases, abstract text, chemical substance and general subject index entries, CAS Registry Numbers, and molecular formulas. Some examples of topics are forensic chemistry, ion exchange, and radiation chemistry, but offerings change with demand.

The *CA Selects* concept has been extended to serve individual organizations. A company may work with *CA* to define a similar product for internal use. Registry Numbers may be used to monitor new information on company products.

Future. CAS is planning to test their substructure searching system during 1980–1981 and to make publicly available the first version of the system by 1982. If successful, it will offer chemists access to the entire registry, now over five million compounds.

CAS Bibliography

A. J. Beach, H. F. Dabek, Jr., and N. L. Hosansky, "Chemical Reaction Information Retrieval from Chemical Abstracts Service Publications and Services," *J. Chem. Inf. Comput. Sci.* **19**(3), 149–155 (1979).

J. E. Blake, V. J. Mathias, and J. Patton, "CA Selects—A Specialized Current Awareness Service," *J. Chem. Inf. Comput. Sci.* **18**(4), 187–190 (1978).

D. L. Dayton, M. J. Fletcher, C. W. Moulton, J. J. Pollock, and A. Zamora, "Comparison of Retrieval Effectiveness of CA Condensates (CA Con) and CA Subject Index Alert (CASIA)," *J. Chem. Inf. Comput. Sci.* **17**(1), 20–28 (1977).

P. G. Dittmar, R. E. Stobaugh, and C. E. Watson, "The Chemical Abstracts Service Chemical Registry System, I. General Design," *J. Chem. Inf. Comput. Sci.* **16**, 111–121 (1976).

R. G. Dunn, W. Fisanick, and A. Zamora, "A Chemical Substructure Search System Based on Chemical Abstracts Nomenclature," *J. Chem. Inf. Comput. Sci.* **17**(4), 212–218 (1977).

M. D. Huffenberger and R. L. Wigington, "Chemical Abstracts Service Approach to Management of Large Data Bases," *J. Chem. Inf. Comput. Sci.* **15**(1), 43–47 (1975).

J. W. Lundeen, "Experimental Use of a Search Profile to Derive a Specialized Information Base from the CAS Data Base," *175th ACS National Meeting, Anaheim, Calif., March 1978.*

I. R. McKinley and A. K. Kent, "ACS/CS Cooperation Agreement," *Chem. Br.* **12**(1), 4–6 (1976).

C. W. Moulton, "Fossil Fuels in Chemical Abstracts," *J. Chem. Inf. Comput. Sci.* **19**(2), 83–86 (1979).

D. C. Myers, J. A. Rathbun, F. A. Tate, and D. W. Weisgerber, "Bridging and Interlinking the Information Resources," *J. Chem. Inf. Comput. Sci.* **16**(1), 16–19 (1976).

R. E. O'Dette, "The CAS Data Base Concept," *J. Chem. Inf. Comput. Sci.* **15**(3), 165–169 (1975).

O. B. Ramsay, "Chemical Abstracts: An Introduction to its Effective Use," *Tape/Slide Audio Course,* American Chemical Society, Washington, D.C., 1979.

L. G. Wade and R. G. M. Cosgrave, "User's Reaction to a Corporate-Designed Current Awareness Bulletin," *Chem. Inf. Comput. Sci.* **22**(3), 179–181 (1980).

M. E. Williams, "Analysis of Terminology in Various CAS Data Files as Access Points for Retrieval," *J. Chem. Inf. Comput. Sci.* **17**(1), 16–20 (1977).

Institute for Scientific Information. The Institute for Scientific Information (ISI) offers a number of information tools for the chemist. The level of complication varies to serve a variety of needs. Most chemists are already familiar with the current-awareness series, *Current Contents.* These reproduce, by permission, the contents pages of core journals in particular disciplines, in weekly, pocket-sized publications for personal scanning. Each issue has a subject index, an index to the principal author of each paper, and an address directory. *Current Contents, Physical and Chemical Sciences* is the primary publication in the series for chemists, but others, such as *Life Sciences* or *Engineering and Technology,* may be of interest.

Current Abstracts of Chemistry (*CAC*), published weekly, provides chemists with a guide to chemical research and technology. It contains abstracts of articles from 107 core journals reporting the synthesis, isolation, and identification of new compounds. Graphic and narrative abstracts are prepared from the original article or selected from the source journal. Extensive use of flow diagrams facilitates rapid scanning, and use-profile and technique-data symbols alert the user to chemical activity of a compound and to analytical procedures used by the investigator (see Fig. 1). The technique-data symbol also highlights articles containing new synthetic methods. Details on these methods appear in the monthly publication, *Current Chemical Reactions* (*CCR*). *CAC* includes reaction schemes, experimental data, bibliographic information, authors' abstracts, and an index section containing journal, author, permuted subject terms, and corporate addresses.

Index Chemicus (*IC*), the companion index to *CAC*, is incorporated into the weekly issues. It contains the following sections: molecular formula, author, subject, biological activity, and an alert to labeled compounds. A list of the journal issues covered each week appears on the back cover. All of the above indexes are cumulated quarterly and annually, the cumulations include a rotaform index of molecular formulas.

All new compounds are encoded using the Wiswesser Line Notation (WLN). These are then cumulated and permuted monthly and annually and are available on microfilm as the *Chemical Substructure Index* (*CSI*).

A personalized current-awareness service that is a product of this family of ISI products is the *Automatic New Structure Alert* (*ANSA*).

Current Chemical Reactions, published monthly, is a guide to new and newly modified reactions and syntheses reported in current journal literature. Flow charts and complete bibliographic information accompany every entry (see Fig. 2). Authors' abstracts and product yields are provided when they are given in the source article. Also included are descriptions of the type of reaction, highlights of techniques used in analyzing compounds, and notices of explosive reactions. Bibliographic information on review articles containing reactions and syntheses is also included. Reports of reactions and syntheses are taken from 107 organic chemistry and pharmaceutical journals. Review articles are drawn from these same source journals plus additional journals and selected books. A list of titles of journals and books covered appears periodically throughout the year in *Current Chemical Reactions*. Each monthly issue of *Current Chemical Reactions* is indexed by author, journal, author affiliation, and permuted subject entry. These four indexes are cumulated annually.

A third family of ISI products is based on their *Science Citation Index* (*SCI*). The printed journal issues bimonthly, with annual and five-year cumulations. The *SCI*, by providing the unique ability to identify papers citing an earlier key paper, allows users to search forward in time. It identifies follow-up work, criticism, work in an interdisciplinary area, and work in a new area with undefined vocabulary. The printed product is in three parts—the *Citation Index*, the *Source Index* (author index), and the *Permuterm Subject Index*. All three of these search approaches are available to users of the on-line version of *SCI*, which is updated weekly and is thus one of the most current of all retrieval resources.

The SDI Service based on the *SCI* is called Automatic Subject Citation Alert (ASCA). The familiar subject or author terms can be included in the search profile, along with cited author or cited reference search keys. ISI also publishes a series of packaged SDIs in narrow areas called ASCATOPICS.

279194

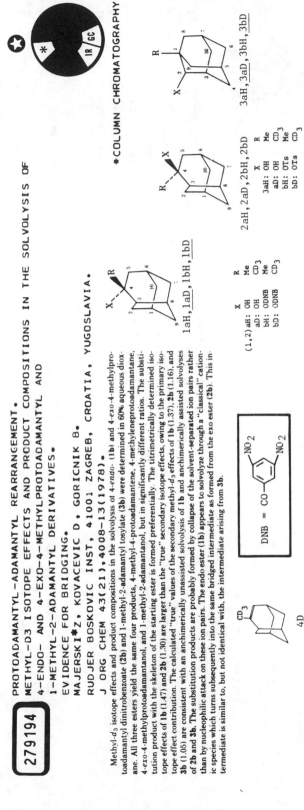

PROTOADAMANTYL-ADAMANTYL REARRANGEMENT.
METHYL-D3 ISOTOPE EFFECTS AND PRODUCT COMPOSITIONS IN THE SOLVOLYSIS OF
4-ENDO- AND 4-EXO-4-METHYLPROTOADAMANTYL AND
1-METHYL-2-ADAMANTYL DERIVATIVES.
EVIDENCE FOR BRIDGING.
MAJERSKI*Z, KOVACEVIC D, GORICNIK B.
RUDJER BOSKOVIC INST, 41001 ZAGREB, CROATIA, YUGOSLAVIA.
J ORG CHEM 43(21),4008-13(1978).

Methyl-d_3 isotope effects and product compositions in the solvolysis of 4-*endo*- (**1b**) and 4-*exo*-4-methylpro-toadamantyl dinitrobenzoate (**2b**) and 1-methyl-2-adamantyl tosylate (**3b**) were determined in 60% aqueous diox-ane. All three esters yield the same four products, 4-methyl-4-protoadamantene, 4-methyleneprotoadamantane, 4-*exo*-4-methylprotoadamantanol, and 1-methyl-2-adamantanol, but in significantly different ratios. The substi-tution product with the skeleton of the starting ester is formed preferentially. The titrimetrically determined iso-tope effects of **1b** (1.47) and **2b** (1.30) are larger than the "true" secondary isotope effects, owing to the primary iso-tope effect contribution. The calculated "true" values of the secondary methyl-d_3 effects of **1b** (1.37), **2b** (1.16), and **3b** (1.05) are consistent with an anchimerically unassisted solvolysis of **1b** and anchimerically assisted solvolyses of **2b** and **3b**. The substitution products are probably formed by collapse of the solvent-separated ion pairs rather than by nucleophilic attack on these ion pairs. The endo ester (**1b**) appears to solvolyze through a "classical" cation-ic species which turns subsequently into the same bridged intermediate as formed from the exo ester (**2b**). This in-termediate is similar to, but not identical with, the intermediate arising from **3b**.

DNB = CO

*COLUMN CHROMATOGRAPHY

Figure 1. Sample abstract from *Current Abstracts of Chemistry* (7). Courtesy of Institute for Scientific Information.

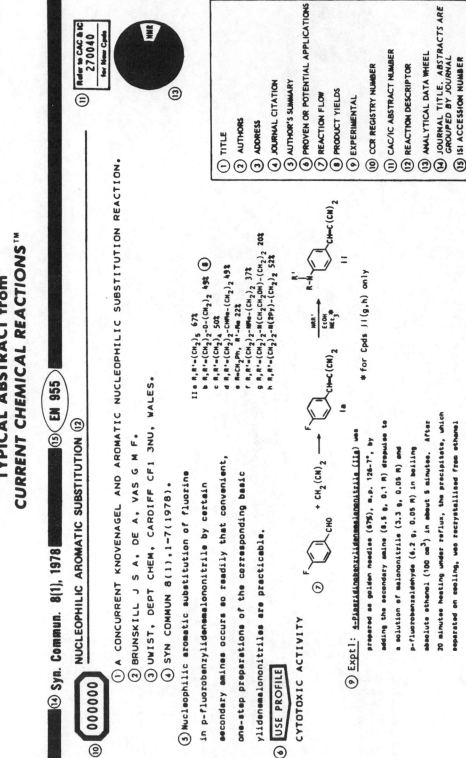

Figure 2. Typical abstract from *Current Chemical Reactions*. Courtesy of Institute for Scientific Information.

There have been many studies using the *SCI* to identify key scientists, core journals, citation networks, and award winners. A sampling of references follows.

P. W. Brennen and W. P. Davey, "Citation Analysis in Literature of Tropical Medicine," *Bull. Med. Libr. Assoc.* **66**(1), 24–30 (1978).
A. E. Cawkell, "Evaluating Scientific Journals with Journal Citation Reports—Case Study in Acoustics," *J. Am. Soc. Inf. Sci.* **29**(1), 41–46 (1978).
R. A. V. Diener, "Transnational Information Flow as Assessed by Citation Analysis," *J. Am. Soc. Inf. Sci.* **30**(2), 114–115 (1979).
P. Ellis, G. Hepburn, and C. Oppenheim, "Studies on Patent Citation Networks," *J. Doc.* **34**(1), 12–20 (1978).
E. Garfield, "Citation Indexing for Studying Science," *Nature* **227,** 669–671 (1970).
E. Garfield and A. E. Cawkell, "Location of Milestone Papers Through Citation Networks," *J. Libr. Hist.* **5**(2), 184–188 (1970).
H. C. Liu, "Faculty Citation and Quality of Graduate Engineering Departments," *Eng. Ed.* **68,** 739–741 (1978).
N. Wade, "Citation Analysis—New Tool for Science Administrators," *Science* **188,** 429–432 (1975).

Patents. In general, patents are an underutilized source of technical information. The significance of this statement has been shown in a study that found that the information in 71% of U.S. patents is never disclosed in the nonpatent literature and that the information content of another 13% is only partially disclosed (8). The patent information that finally is published in journals generally issues several years after the patent (see also Patents, literature).

Patent literature affects research and development in three areas as: a fruitful source of information regarding the economic potential of existing or near future art; a teacher of the art; and a guide to application areas for marketing activities (9). An extension of this commentary is the ability to mine the richness of patent information data bases as intelligence tools. For instance, patent information can be used to predict trends or to analyze competitors' research and development in a given area (10–11).

The immense volume of patent literature may be intimidating. There are 100,000–150,000 basic (first disclosed) chemical patents issued annually throughout the world, and 150,000 equivalent chemical patents (12). But patents are technical publications: once the reader is comfortable with their format and jargon (13), patents have more experimental detail than most journal articles (14).

To use patent information effectively generally requires the assistance of experienced information professionals. Each service described below approaches a search from a different perspective. Searching caveats are based on the following: each service has different rules of coverage (subject and country); each uses a different depth of the U.S. or International Patent Classifications; each may index end-product chemicals only or include reactants and intermediates; and each may or may not enhance the titles. References 12 and 15–17 provide in-depth discussions of these differences.

A new journal, *World Patent Information,* began publication in 1979 (K. G. Saur Verlag). For those interested in patent policy, a collection of papers from a recent ACS symposium has been published (18), and the journal *Research Management* frequently discusses the issue.

National Bulletins. Individual national patent offices, and now the European Patent Office (EPO) as well, issue bulletins disclosing the title, assignee, and abstract of approved applications. In the United States, the *Official Gazette,* published weekly on Tuesdays, is organized by classification. About 30% of the patents reported are chemical (13).

Derwent.   Derwent Publications, Ltd. (19) covers patents issued in 24 countries, plus EPO patents and international Patent Cooperation Treaty (PCT) patents. Some of these are from slow-issue countries, which examine applications, and some are from fast-publishing ones. Derwent's special value is the comprehensiveness and timeliness of its service. Consequently, many large industrial companies subscribe to its services. Derwent *World Patents Index* (*WPI*) contains information on all chemical areas since 1970. For pharmaceutical, agricultural, and polymer-related chemicals, the service extends back to different points in the 1960s.

For keeping abreast of new inventions, Derwent targets two bulletins for chemists. Twelve *Alerting Bulletins—Classified* provide very rapid notice of disclosures with brief abstracts of all basic patents and all examined equivalents. The *Profile Booklets* are a few weeks behind the *Alerting Bulletins* but include longer abstracts and cover more defined subject areas. Two other bulletin types used by patent professionals and chemists are the *Alerting Bulletins—Country Order* and the *Basic Abstract Journal.*

Derwent has other alerting services designed for patent attorneys and numerous searching products—manual, microform, and computerized. In addition to complex subject-coding techniques, access by title word, country, patentee, classification, and patent family is possible. Although bibliographic searching is excellent, subject searchability varies. There are now two sources for on-line access: SDC in the United States and INFOLINE, a UK host on the EURONET–DIANE System. For a fee, Derwent's Technical Services Division performs searches for subscribers.

Chemical Abstracts.   *Chemical Abstracts* does not treat patents as legal documents, so their coverage is selective; to be covered a patent must disclose new chemical information. In some countries, *CA* covers all basics; in others, only patents issued to nationals. But CA's chemical indexing is not restricted to end products. It indexes all substances, uses, and methods of manufacture included in the claims and additional uses or substances if actually prepared (12,17,20–21).

IFI–Plenum.   IFI–Plenum has several levels of patent-retrieval services. The most readily accessible is the CLAIMS family of data bases from the Lockheed-DIALOG on-line service. Together these cover an extended time period but the indexing is shallow.

CLAIMS–CHEM.   1950–1970 bibliographic information, including U.S. classification codes, for United States chemical patents. Revised annually to accommodate U.S. Patent and Trademark Office (PTO) reclassification.

CLAIMS–U.S. PATENTS.   1971–1977 bibliographic information as above but for all U.S. patents.

CLAIMS–U.S. PATENT ABSTRACTS.   Claims from the *Official Gazette* are now included, with chemical structures in linear form and the presence of mechanical drawings noted; covers the time period 1978 to the present.

CLAIMS–U.S. PATENT ABSTRACTS WEEKLY.   Unedited, rapid information, for use until the monthly edited file is added to CLAIMS–U.S. PATENT ABSTRACTS.

CLAIMS–CLASS.   A keyword index to the U.S. PTO classification manual.

IFI–Plenum also has a search service for use of their COMPREHENSIVE DATA BASE, which adds deep uniterm indexing and chemical fragmentation coding (from DuPont) to the data base (starting 1950). In addition, they publish a *UNITERM Index,* now available on line, and the *IFI Assignee List.*

APIPAT. The American Petroleum Institute Patent data base (APIPAT), dating from 1964, is available on SDC (22). Although its scope is relatively narrow, the indexing is done from a controlled vocabulary. This vocabulary provides a simple fragment or chemical-aspects system that minimizes false retrieval by linkages of the aspects while allowing for some structure searching. The use of role indicators allows a chemical to be searched as a starting material or as an end product.

In conjunction with API's retrospective data base, a number of patent alerts are produced in cooperation with Derwent for use by the scientists in API's subscribing organizations.

INPADOC. INPADOC is a joint venture produced by the Austrian government and the World Intellectual Property Organization (WIPO). It obtains, standardizes, and compiles the bibliographic details of patents from 46 national patent offices and two international organizations. The data base contains 7.5 million patents dating from 1968. It can be used to locate individual inventors or corporate patentees. However, subject information can be obtained only through the broad and imprecise medium of international patent classes, with only the patent title as an indication of the patent's content. There are no abstracts. One of INPADOC's prime uses is for keeping track of equivalent members of a patent family in countries around the world. The INPADOC products are marketed in the United States by IFI–Plenum Data Corporation. In 1980, the most recent six-weeks of INPADOC became a searchable file on DIALOG.

Search Check. Search Check, an information service based in Arlington, Virginia, applies the concept of citation searching to patents. Every U.S. patent issued since 1947, and the patents cited therein, are in the Search Check data base. This service is very helpful for locating references that the patent examiners cite. At this time, Search Check runs the computer search for customers on a one-time or contract basis.

Bibliography

H. W. Grace, *A Handbook on Patents,* Charles Knight & Co., Ltd., London, Eng., 1971.
S. M. Kaback, "A User's Experience with the Derwent Patent Files," *J. Chem. Inf. Comput. Sci.* 17(3), 143–148 (1977).
E. J. Saxl, "The Significance of a Patent," *Am. Lab.* 10(9), 31–35 (1979).
H. Skolnick, "Historical Aspects of Patent Systems," *J. Chem. Inf. Comput. Sci.* 17(3), 114–121 (1977).
J. T. Maynard, *Understanding Chemical Patents: A Guide for the Inventor,* American Chemical Society, Washington, D.C., 1978.
J. F. Mezieres, "The European Patent—Henceforth a Reality," *Chemtech* 8(11), 658–661 (1978).
F. Newby, *How to Find Out About Patents,* Pergamon Press, New York, 1967.
C. Oppenheim, "Recent Changes in Patent Law and Their Implication for Information Services and Information Scientists," *J. Doc.* 34(3), 217–219 (1978).
W. Pilch and W. Wratschko, "INPADOC: A Computerized Patent Documentation System," *J. Chem. Inf. Comput. Sci.* 18(2), 69–75 (1978).
E. S. Turner, "Patent Literature," in A. Standen, ed., *Kirk-Othmer Encyclopedia of Chemical Technology,* 2nd ed., Vol. 14, Wiley-Interscience, New York, 1967, pp. 583–635 (see also Vol. 16, 3rd ed.).

The NIH-EPA Chemical Information System. The still developing Chemical Information System (CIS) of the National Institutes of Health and the Environmental Protection Agency (NIH–EPA) is an excellent example of computer utilization to help the chemist in the laboratory. Several United States agencies were motivated to use computers to assist both the chemist and the regulator in identifying substances, solving problems, relating chemical structure to biological or environmental behavior, and responding to emergencies. CIS is a key tool for the Interagency Regulatory Liaison Group (IRLG), consisting of the EPA, FDA, OSHA, and the Consumer Product Safety Commission. The system is fully interactive, has referral, identification, characterization, calculation, and graphics features, and contains chemical, biological, and bibliographic data bases.

The Structure and Nomenclature Search System (SANSS) is the heart of CIS (see Fig. 3) (23). SANSS works on a unified data base, built from many files that identify chemicals of commercial importance or known biological activity (ie, drugs, pesticides, commodity chemicals, food additives), and therefore relevant to these regulatory agencies (see Table 1). The CAS Registry Number is used as the unifier for the various files. Reported or regulated substances that are structurally undefined get pseudo-Registry Numbers. Ultimately, because of the overlap in files, and because of the limited number of chemicals actually used commercially, the file is expected to total 175,000–200,000 substances.

The search techniques of SANSS employ the following data elements to identify particular compounds: nomenclature, ring, fragment, molecular formula, substructure, and full structure.

The display options are by chemical structure: *CAS Collective Index Names;* synonyms, common or trade names; molecular formulas; and list of files containing the substance. One can also retrieve by CAS Registry Number. Once Registry Numbers are located, one can search the other CIS data bases. Conversely, the chemist can use the mass-spectral system to identify an unknown, and use the Registry Number in SANSS to retrieve the structure.

Among the chemical-data components currently available in CIS are the Mass Spectral Search System, X-Ray Crystallographic Search System, X-Ray Single Crystal Search System, and Carbon-13 Nuclear Magnetic Resonance Spectral Search System. Examples of the biologically oriented components are *Registry of Toxic Effects of Chemical Substances,* Ames Test, and AQUATOX. Files containing *Federal Register* notices and mass-spectrometry bulletins are bibliographic. One last component category is the Mathematical Modeling Laboratory, where the data from other components are used to test predictive theories of structure–activity relationships.

Bibliography

G. W. A. Milne and S. R. Heller, "The MSDC/EPA/NIH Mass Spectral Search System," *Am. Lab.* 8(9), 43–54 (1976).
S. R. Heller, G. W. A. Milne, and R. J. Feldman, "Quality Control of Chemical Data Bases," *J. Chem. Inf. Comput. Sci.* 16(4), 232–233 (1976).
R. J. Feldmann, G. W. A. Milne, S. R. Heller, A. Fein, J. A. Miller, and B. Koch, "An Interactive Substructure Search System," *J. Chem. Inf. Comput. Sci.* 17(3), 157–163 (1977).
G. W. A. Milne and S. R. Heller, "The NIH–EPA Chemical Information System," in D. H. Smith, ed., *Computer-Assisted Structure Elucidation, ACS Symposium Series 54,* American Chemical Society, Washington, D.C., 1977, pp. 26–45.
S. R. Heller and G. W. A. Milne, "The NIH–EPA Chemical Information System," in W. J. Howe, M. M. Milne, and A. F. Pennell, eds., *Retrieval of Medicinal Chemical Information, ACS Symposium Series 84,* American Chemical Society, Washington, D.C., 1978, pp. 144–167.
S. R. Heller, G. W. A. Milne, and R. J. Feldmann, "A Computer-Based Chemical Information System," *Science* 195, 253–259 (1977).
G. W. A. Milne, S. R. Heller, A. E. Fein, E. F. Frees, R. G. Marquart, J. A. McGill, J. A. Miller, and D. S. Spiers, "The NIH–EPA Structure and Nomenclature Search System," *J. Chem. Inf. Comput. Sci.* 18(4), 181–186 (1978).

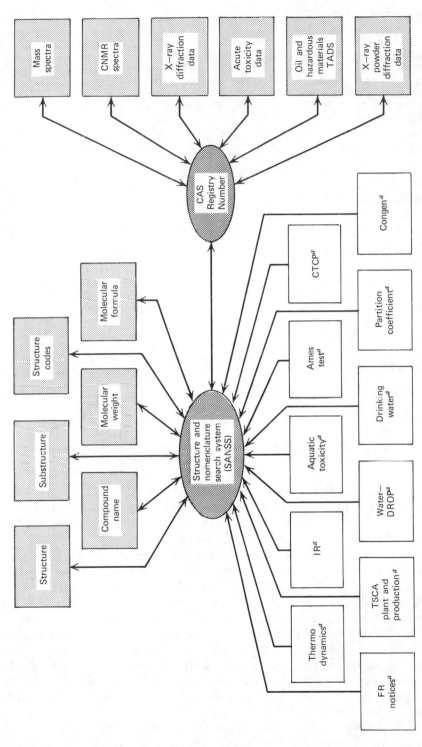

Figure 3. The NIH–EPA Chemical Information System (23). [a] Under development. Courtesy of the American Chemical Society.

Table 1.   SANSS Files [a]

SANSS file	Substances	SANSS file	Substances
TSCA inventory	43,278	NBS x-ray crystal	18,338
CIS mass spectrometry	25,560	EPA effluent guidelines	125
CIS carbon-13 nmr spectrometry	3,805	EPA organic chemical producers	375
EPA pesticides active ingredients	1,453	IPC chemical product	104
EPA OHM/TADS	858	IPC chemical plant	103
Cambridge x-ray crystal	14,854	NSF chemicals list	225
Merck Index	8,959	EROICA thermodynamics	4,492
EPA pesticides analytical reference standards	473	PHS149 carcinogenic activity	4,447
EPA STORET	234	NIOSH RTECS	19,882
EPA chemical spills	577	NIOSH NOHS	4,560
EPA AEROS SOTDAT	572	ORNL EMIC	4,030
NIMH psychotropic drugs	2,039	ORNL ETIC	3,241
EPA AEROS SAROAD	65	EPA selected organic air pollutants	579
NBS proton affinities	440	Clean Air Act Section 112	4
CPCS CHEMRIC	890	EPA/NCTR study (1976)	91
EPA pesticides registered inert ingredients	735	EPA environmental carcinogen assessment program	21
NBS gaseous ions	3,167	EPA restricted use pesticides	23
NFPA hazardous chemicals	396	EPA compounds for mutagenicity evaluation	25
FDA/EPA pesticides reference standards	613	CIIT priority chemicals lists (toxicological)	27
U.S. International Trade Commission	9,193	NMFS survey of trace elements	15

[a] Ref. 23.

S. R. Heller and G. W. A. Milne, "The NIH–EPA Chemical Information System," *Environ. Sci. Technol.* **13**(7), 798–803 (1979).

S. R. Heller and G. W. A. Milne, "Chemcorner," *Database* **2**(3), 69–79 (1979).

H. J. Bernstein and L. C. Andrews, "The NIH–EPA Chemical Information System," *Database* **2**(1), 35–49 (1979).

Other Bibliographic Data Bases.   There are many data bases of interest to chemists beyond those directed to chemistry. Data bases available in the United States are listed in ref. 24. The number of data bases listed is overwhelming, however, the quality and variety should encourage chemists to use these resources.

Careful selection of data bases, based on the needs of each particular query, can save time. For example, if a user is looking for an overview of an area or an issue or wants to sense the amount of activity in a field, searching the mission-oriented data bases, such as ENVIROLINE or ASFA (Aquatic Sciences and Fisheries Abstracts), may be more efficient than searching the core displinary data bases such as CA SEARCH or BIOSIS PREVIEWS (on-line versions of *Chemical Abstracts* and *Biological Abstracts,* respectively). On the other hand, to achieve any degree of comprehensiveness, multidata-base searching is necessary. Despite some overlap, the indexing perspective of each service varies and so retrieval varies. When many files are searched, overlap is annoying. Techniques to eliminate redundancy in the final result now are being developed.

Other systems to combat the confusion resulting from the richness of the available data bases are in use or are being evaluated. Selector systems allow identification of the data bases that contain terms or parameters relevant to the search. Vocabularyswitching systems translate the original search terms (search strategy) into the language of other data bases.

On-line Chemical Dictionary Files.   On-line chemical dictionary data bases, companions to both bibliographic and numerical data bases, are not traditional dictionaries at all. They do not define chemicals but are designed to assist chemists in identifying CAS Registry Numbers and all the names and synonyms of chemicals. They are primarily identifier or locator tools with some substructure-searching capability. They thus enhance comprehensive searching in other data bases. Because the four dictionary files now available can be expected to change in scope, and because others surely will be developed soon, the following discussion of features is general. The file names, in the order they became available, are CHEMLINE of NLM, CHEMNAME of Lockheed, SANSS of NIH–EPA, and CHEMDEX of SDC. The Lockheed, NLM, and SDC chemical dictionaries are based on information from the *Chemical Abstracts Registry Nomenclature File,* which each search service has enhanced beyond the straightforward Registry Number, name (8th or 9th collective index), synonyms, trade names, molecular formulae, and cross-reference Registry Number. One can do a complex substructure search based on name fragments, such as dichloro or di or chloro, or by molecular formula and fragment, including element count or element presence. Searching is by position in the periodic table, ring counts, ring sizes, ring-element analysis, and even keyword or category (ie, polymer, alloy). CHEMLINE adds the Wiswesser Line Notation and pointers that specify the presence of the chemical in other NLM data bases.

Bibliography

R. E. Buntrock, "Chemcorner," *Database* **2**(1), 33–34 (1979).
S. R. Heller and G. W. A. Milne, "Chemcorner," *Database* **2**(3), 71 (1979).
J. Kasperko, "Online Chemical Dictionaries," *Database* **2**(3), 24–35 (1979).
P. E. Pothier, "Substructure Searching in CHEMLINE," *Online (Weston, Conn.)* **1**(2), 23–25 (1977).
R. J. Schultheisz, D. F. Walker, and K. L. Kannan, "Design and Implementation of an On-line Chemical Dictionary (CHEMLINE)," *J. Am. Soc. Inf. Sci.* **29**(4), 173–179 (1978).

Nonbibliographic Data Bases.   The newest aspect of the information explosion is the increasing number of data bases that contain numerical information such as physical properties, product consumption, or social statistical data. These are not only fact-finding systems but also tools for manipulating facts for analyses or evaluation. Users can even add their own data and merge or compare it with that in the data base. Unfortunately, most nonbibliographic data bases are in different search-service systems from the bibliographic ones. Thus the software is different. But over the next few years there will be more numerical data bases, they will be larger, and more will contain evaluated data. Cuadra Associates, Inc., publishes a comprehensive and regularly updated list of numerical data bases (25).

These information resources are intended for use in everyday work. The opportunity to interact directly with nonbibliographic systems is available to the individual chemist. The chemical-information specialist will continue to use these systems in an intermediary role, and to act as informer, trainer, and coordinator for others' use of nonbibliographic services.

Thus, the tremendous wealth of scientific and technical information contained in the ever increasing number of information resources can be seen. This proliferation of literature increases the likehood of not finding the information needed. Don't be concerned. Consult a librarian or information specialist. These information professionals can help in a search or obtain pertinent information. Although information is becoming more obscure to end users it is, at the same time, becoming more focused or specialized. The evidence of this is the growth and development of business information sources useful to chemists and chemical technologists.

BUSINESS INFORMATION

Although the nature of technical information is fundamentally universal, economic information remains essentially national and is often private or sensitive as well. Indeed, each country conducts its own affairs according to its individual history, traditions, social organization, political environment, and particular attitudes. The result is that this information is confined and subject to local constraints and, therefore, when accessible, is available in a variety of forms.

National authorities, however, as well as regional or international organizations, have become increasingly aware of the importance of economic information. Although it is vital to private, regional, and national economic development, it remains scattered, sometimes insufficiently known, yet abundant. The many sources of economic information have been the object of various surveys but no satisfactory effort has been undertaken yet to organize it suitably.

To conduct and manage its business, the chemical industry requires vast amounts of primary information to be selected, analyzed, organized, and transformed for use by management. The principal areas of important nontechnical information are the political, social, legal, and economic environments; past, present, and future markets; competition; and finances, prices, taxes, and tariffs.

The Political, Social, Legal, and Economic Environments.   Every country presents a unique environment for the conduct of business. Attitudes toward chemical business, whether national or foreign, differ among countries and change with time. Political and social aspects that can influence chemical operations include governmental stability, its current and anticipated attitudes toward both national and foreign business, the possibility that it will enter into a specific chemical business through part ownership or nationalization, and ways in which it might favor local manufacturing industries rather than foreign companies or enterprises. Duties, taxes, port charges, import and export restrictions, and price regulations are factors essential to any contemplated or operating business. Labor practices and laws, trade union attitudes, levels and trends of unemployment, as well as the outlook for inflation and foreseeable levels of wage increases require careful consideration. Recent, potential, and long-range growth of the economy in general and in specific businesses, corresponding consumption, balance of payments, and monetary policies are crucial.

Markets.   The understanding of markets whether past, present, or future, requires a comprehensive body of information. This information includes the list of all significant products that are manufactured, imported, exported, sold, and potentially pertinent to the business. Product information includes forms, grades, types, and specifications as well as sales, prices, and regulations concerning storage, transportation, labeling, and environmental aspects. Materials that have been or could be replacements for products also need to be known. This in turn calls attention to factors that bear on the possible vulnerability of any one product to price, local restrictions or new regulations, public health, or public opinion.

An understanding of the market requires knowing the supply and demand for each individual product or group of products within each geographical area contemplated for manufacturing, importing, and trading. This information covers the current year and past periods of five to ten or more years depending on particular products and situations. It is expressed in terms of physical volume and prices. Yearly, quarterly, or monthly figures may be required depending on seasonal variations that may affect forecasts.

Likewise, the historical and forecasted demands for products by main users are important because they may reveal characteristics that indicate whether the demand is cyclic and price-sensitive. They may reveal that the end-use product is vulnerable to imports or other products manufactured at lower cost.

Forecasting demand requires various kinds of data, eg, product supply and demand, end-use breakdown, consuming industries, and overall economic growth. These data should go back as far as possible; it should, by a rule of thumb, cover a period at least twice as long as the period of the forecast.

Forecasting accounts for supplies from domestic sources, imports from foreign countries, and exports to other regions. This means that current and expected production capacities compared to imports in countries that provide export outlets for domestic manufacture must be thoroughly reviewed. A detailed demand-and-supply analysis in terms of major manufacturers, newcomers, importers, consuming industries, and customers requires further information, which is described below.

Companies, Manufacturers, and Customers. A company's success depends upon careful, objective analysis of the industry and end-use markets and trends, competition, significant manufacturing and technical trends, the projected availability of feedstocks, and energy requirements. Therefore, company information must include financial performance and manufacturing capabilities and capacities. In this respect, the technical-business analyst requires reliable data on current, historical, and projected production capacities by product. Data compilations regarding specific plants must include information such as owning companies, geographical location, licensed processes, engineering design, contractor who built the plant, start-up date, and dates of debottlenecking, expansions, and shutdowns. General industrial conditions that influence prices must be reviewed. These include new suppliers, technological changes, increases in average industrial-plant size, required capital expenditures for new capacities, and likely increases of capital with time. General industrial conditions are indicative not only of current and likely utilization rates but also of trends relating to manufacturers securing their own captive outlets or to customers providing their own source of supply. The position, strengths, and weaknesses of companies are measured in financial terms such as sales and profit trends, capital expenditure, cash flow, interest levels, return on employed capital, and value added per employee.

The Different Aspects of Information. The data requirements for technical-business information can be succinctly summarized as *macroeconomic:* world, region, country, and area—economic, monetary, tariffs, taxes, controls, commerce prices, cost, gross national product (GNP), gross domestic product (GDP), population, movement of capital, and employment; and *sectorial:* company, supplier, customer, country, government—production, industrial indexes, technology, markets, distribution, organization, and structure.

Sources

Political, Social, Legal, and Economic Environments. Because of national differences every country has a unique set of information sources on the economic, legal, social, and political environment. Access to these sources is easy in countries such as the United States or the countries of the European Economic Community; elsewhere it is much more difficult; in countries with centrally planned economies, such information is minimally accessible.

Therefore, a first resource is the international and regional organizations that have systematically collected and organized data and information over several years. Organizations like the United Nations (UN), the Organization for Economic Cooperation and Development (OECD), the European Economic Community (EEC), and the International Monetary Fund (IMF) issue valuable publications for the technical-business analyst. These publications include monographs, specialized periodicals, statistical yearbooks, as well as monthly and quarterly statistical bulletins and conference proceedings.

Information about international organizations appear in publications such as refs. 26–27. The latter provides not only detailed information on international organizations, institutions, or associations but also basic information on individual countries and purely national organizations whether governmental, political, industrial, judicial, or educational. The information for each country includes general and statistical surveys of its recent history together with an overview of the economy. Also included is a directory of the diplomatic corps, the press and publishers, finance, banks, trade unions, chambers of commerce, and universities.

All international organizations publish enormous amounts of material that may not be fully covered in any one bibliography. Although each organization issues various indexes and catalogues, an overview of all that is available in any specific field is difficult. Some useful references, source guides, and information-providing services are listed below.

United Nations.   Addresses: United Nations Publications, Palais des Nations, 1211 Geneva 10, Switzerland; United Nations Publications, Sales Section, Room A-3315, New York, N.Y. 10017, U.S.

A Guide to the Use of United Nations Documents Including Reference to the Specialized Agencies and Special UN Bodies, B. Brimmer and co-workers, Oceana Publications, Dobbs Ferry, New York, 1962, is an excellent key to the literature published by the UN. Likewise, *UNDOC Current Index: United Nations Document Index*, Dag Hammerskjold Library, United Nations, New York, published bimonthly, the *United Nations Document Index (UNDI)*, Cumulated Index to Vol. 1–13, KTO Press, Millwood, New York, 1974, and *United Nations Publications in Print 1979–1980*, United Nations Sales Section, New York, *The United Nations Document Index*, and the *Catalogue: United National Publications* are useful for locating and learning the contents of publications that can be purchased from the UN.

The publications of the United Nations Statistical Office pertinent to the chemical sector in particular are: *Demographic Yearbook; Labour Supply and Migration* (by regions); *World Trade Annual; Yearbook of National Accounts Statistics; Statistical Yearbook; Monthly Bulletin of Statistics;* and *The Growth of World Industry* (published every two to three years).

The Department of Economic and Social Affairs publishes *World Economic Survey* (annual).

Each regional economic commission issues it own publications. The Economic Commission for Europe publishes *Economic Survey of Europe* (annual); *Economic Bulletin of Europe* (annual); and *Structure and Change in European Industry, 1977.*

The Economic Commission for Asia and the Far East publishes the *Statistical Yearbook for Asia and the Far East.* The Economic Commission for Latin America publishes the *Analyses on Projections of Economic Development.* The Economic Commission for Africa publishes the *Foreign Trade Statistics of Africa* and the *African Economic Indicators.* The UN also publishes an annual series of *Studies on Selected Development Problems in Various Countries in the Middle East.*

The International Labour Organization (ILO). Address: 1211 Geneva 22, Switzerland. The International Labour Organization (ILO), a specialized United Nations agency, deals with the improvement of living and working conditions of the member countries. ILO issues its *Catalogue of ILO Publications* irregularly. Information on new publications can be found in the *International Labour Review.* The publications of ILO that are most pertinent for industry are: *Yearbook of Labour Statistics; Bulletin of Labour Statistics* (quarterly); and *Social and Labour Bulletin* (quarterly). *The Social and Labour Bulletin* provides information on such topics as social policy, labor legislation, labor relations, collective agreements, employment, working conditions, wages, social security, equal opportunity, multinational enterprises, and trade unions on a country-by-country basis. ILO also publishes a series of special reports on topics such as productivity and workers' involvement in management.

The Food and Agricultural Organization (FAO). Address: Viale della Tezme di Caracalla, 00100 Rome, Italy. This UN agency issues considerable amount of economic information on food, agriculture, fisheries, and forestry; this information is referenced in the *FAO Documentation: Current Index* and the *Catalogue of FAO Publications.* Important FAO statistical publications include the following: *FAO Trade Yearbook; FAO Production Yearbook;* and *Monthly Bulletin of Agricultural Economics and Statistics.* FAO also issues useful reviews and bulletins: *FAO Commodity Review and Outlook; Annual Fertilizer Review; Per Caput Fibre Consumption;* and *Commodity Bulletin Series: Impact of Synthetics on Jute and Allied Fibres.*

The International Monetary Fund (IMF). Address: 19th and H Streets, N.W., Washington, D.C. 20431, U.S. The IMF issues several publications for the corporate financial expert. Information on foreign-exchange-rate systems for individual countries appears in *Annual Report: Exchange Arrangements and Exchange Restrictions.* International statistics on all aspects of domestic and international finance appears in *International Financial Statistics Yearbook,* which provides annual data for the 1949–1978 period. IMF also publishes a *Balance of Payments Yearbook* and a quarterly entitled *Finance and Development.*

The General Agreement on Tariffs and Trade (GATT). Address: Villa le Bocage, Palais des Nations, 1211 Geneva 10, Switzerland. Trade information on Eastern European countries that are members of the COMECON (Council of Mutual Economic Assistance, also known as CMEA) is provided by GATT in the annual publication, *International Trade,* which generally covers the world.

The Organization for Economic Cooperation and Development (OECD). Address: 2, Rue André-Pascal, 75 775 Paris Cedex 16, France. Every two to three years, the Organization for Economic Cooperation and Development issues *Catalogue of Publications,* which is updated in the interim by supplements. Because the primary objective of the OECD is to encourage the trade, economic growth, and improvement of living standards of the member countries (ie, Western Europe, Yugoslavia, Canada, Japan, the United States, Australia, and New Zealand), this organization publishes several items that bear on economists' and business analysts' activities. OECD's Department of Economics and Statistics produces an impressive number of statistical tables and graphs each year. Its major publications are:

Industrial Production, Historical Statistics (1955–1971; 1960–1975).

Indicators of Industrial Activity (quarterly), which has been published since 1979 and replaces the former *Industrial Production* (quarterly supplement) and the *Short Term Economic Indicators for Manufacturing Industry,* is jointly issued by the Industry Division and the Economic Statistics and National Accounts Division and surveys the economic development of industries. It is divided into two sections: statistical data and qualitative business-trends. *Main Economic Indicators, Historical Statistics* (1955–1971; 1960–1975) provides monthly, quarterly, and annual data on gross national product, gross domestic product, industrial production, trade, unemployment, money supply, savings, consumer prices, consumers' expenditures, governmental expenditure, and the like.

Main Economic Indicators (monthly) contains data on recent economic developments of member countries, the long-term outlooks, and is divided into three sections: indicators by subject, indicators by country, and price indexes (data are presented both in graphical and tabular form).

OECD Economic Outlook (July, December) reviews and analyzes economic trends and short-term forecasts that bear on the world's economy (supplemented by occasional studies on topics such as capital movements, income distribution, international competition, and the like).

OECD Economic Surveys are annual economic surveys for each member country that analyze general trends in demand and activity, balance of payments, and economic policy in fiscal, monetary, employment, and industrial terms; the short-term outlook also provides international comparisons.

National Accounts of OECD Countries are published in two separate volumes: the first gives for each country the main aggregates in the form of graphs, the growth in real terms and implicit price deflators, the gross domestic product, and its main components from 1960 to 1975, the main aggregates in national currencies from 1950 to 1975 as well as main components of final expenditure, provides comparative tables of national accounts in U.S. dollars, volume and price indexes, population, and exchange rates; the second volume gives detailed statistics for the last twelve years on domestic product and consumption, gross domestic product, composition of gross capital formation, income, and outlay transactions of government and households, financing of gross capital formation, external transactions, and distribution of national disposable income.

OECD Financial Statistics (June, December) is a financial tool that assembles large amounts of data from individual national publications in different languages, provides information on capital operations, financial transactions, interest rates, markets for long-term securities, lending and borrowing operations of groups of financial institutions, governmental finances, transactions of each economy with the rest of the world, rates of the European currency market, international bond issues, and security issues.

The Chemical Industry (yearly) breaks down the chemical industry according to the Standard International Trade Classification (SITC) and provides production indexes and trends for the major branches of the chemical industry, turnover, value added and investment data, labor, price indexes, trade figures, and production and consumption data.

The OECD also issues, under the title *Energy Balances of OECD Countries* (1960–1974; 1974–1976), statistical data on energy that is conveniently expressed in common units allowing total-energy estimation, forecasting, and studies of conservation and substitution. The balances are calculated largely on the basis of data published in the following:

Statistics of Energy (yearly).

Oil Statistics (yearly), which provides detailed statistics on the supply and disposal of crude oil and natural gas, on feedstocks, and their major derivatives, on production and consumption by end-use sectors, on source of imports, and on destinations of exports.

Quarterly Oil Statistics, which provides timely and detailed statistics on oil supply and demand, ie, balances of production, trade, refinery throughput, final consumption, stock levels, and trade data.

Trade by Commodities—Market Summaries: Imports and *Trade by Commodities—Market Summaries: Exports,* which give foreign trade statistics covering all goods with indication of imports' origin and exports' destination. For any individual SITC code, values are in U.S. dollars, and quantities are in metric units.

The European Economic Community (EEC).   Because the EEC has not selected a unique location for its headquarters (they operate from Brussels, Luxembourg, and Strasbourg), locating documents released by any one of the principal organs is difficult. The principal publishing groups are the Commission of the European Communities (CEC), the Council of Ministers, the European Parliament or Assembly, the Court of Justice, The European Coal and Steel Community (ECSC), The European Atomic Energy Community (Euratom), and the European Investment Bank. The addresses of EEC information offices in member countries and in some nonmember countries such as Japan and the United States, are listed in the *Europa Yearbook.* These information offices provide valuable assistance and are an excellent starting point.

EEC publications are issued in all of the official languages of the Community: Danish, Dutch, English, French, German, and Italian. The list of these publications appears in the *Publications of the European Communities: Catalogue* and in *Official Publications,* issued by the information offices in the member countries.

The main periodicals published by the Commission of the European Communities are the following: *Official Journal of the European Communities: Series L* (Legislation), which gives community secondary legislation coming into force (at least daily); *Series C* (Communications), which gives details on draft secondary legislation, information of general interest, summaries of the proceedings of the European Assembly and the Court of Justice (almost daily); and *Bulletin of the European Communities* (monthly), issued jointly by the ECSC, EEC, and Euratom. This bulletin contains three parts: (*1*) special features; (*2*) current activities such as economic and monetary policies, internal market and industrial affairs, competition, finance and taxation, employment, customs, environment, agriculture, energy, transport, research, and development management information; (*3*) documentation on European units of account, additional references in the *Official Journal,* infringement procedures, public opinion in the Community, publications of the European Communities. *European*

Economy (March, July, November) gives short-term economic trends and prospects with reports and studies on problems of current interest, reviews on the economies of each member state, and a statistical index containing the main economic indicators. *European Economy—Supplements* (monthly): *Series A* gives recent economic trends with tables and graphs; *Series B* gives business-survey results; *Series C* gives consumer-survey results. *Graphs and Notes of the Economic Situation in the Community* (monthly) gives data and graphs on industrial production, trade, employment, consumer and wholesale prices, governmental budgets, long-term interest rates, and the like. *Report of the Results of the Business Surveys* details results of surveys carried out among heads of enterprises in the community (three per year).

The Statistical Office of the European Communities publishes many series on behalf of the different organs of the Community. *Eurostatistics* (monthly) provides data for short-term economic analyses on national accounts, population, employment, output of products, trade, prices, balance of payments, industrial production, opinions in industry, and financial statistics. *Eurostat Industrial Short-Term Trends* (monthly) provides graphs and data on production, turnover, new orders, number of employees, wages and salaries, by country for key sectors and products of the economy. *Eurostat National Accounts ESA, Aggregates* (yearly) provides historical data covering fifteen or more years, in the form of comparative tables and country tables, and in terms of gross domestic product and market prices, compensation of employees, consumption of fixed capital, net operating surplus of the economy, net national disposable income, net national savings, lending and borrowing, final domestic uses, consumption, capital formation, trade, price indexes, population, and employment. *Eurostat Energy Statistics Yearbook* gives a detailed survey of the energy economics for the nine-country Community, the six-country Community, and for each member country over a period of ten to sixteen years. Its quarterly bulletin provides data on the overall energy balance-sheet for main items of energy supply for the whole community and for each member country; it provides monthly data for each energy source. *Iron and Steel Yearbook and Iron and Steel Bulletin* (bimonthly) provide data on production, employment and wages, orders and deliveries, and trade. *Eurostat Industrial Statistics Yearbook* provides data on production, employment and wages, orders and deliveries, and trade and provides historical series of industrial-production indexes back to 1968 and data on the production of basic materials and manufactured goods covering approximately 500 items back to 1958. *Eurostat Analytical Tables of Foreign Trade NIMEX* (yearly) represents a statistical breakdown of the Community's foreign trade based on the EEC's Common Customs Tariff (CCT), which was derived from the 1955 Brussels Tariff Nomenclature. NIMEX (Nomenclature of Imports and Exports) has about 6500 headings; a CCT-NIMEX correlation is provided. *Eurostat Monthly External Trade Bulletin* provides data for broader categories of products with an indication of trends in trade of the EEC countries. *ACP: Statistical Yearbook* is a selection of the main demographic, economic, and social indicators relating to the ACP countries, ie, those that have signed the Lome Convention. The data, which extend back five to ten years, cover population, national accounts, transport, trade, external aid, food supply, finance, balance of payments, and agricultural and industrial production.

Industrial Associations and Confederations.   In a number of countries, companies in the same industrial sector are often members of trade associations or employers' organizations. These associations publish much valuable information in the form of periodical bulletins, occasional publications, reports, yearbooks, directories, and statistical data, all available to member associations and companies. These organizations and the information they release can provide good insight into the individual countries' developments, trends, and industrial sectors.

Most associations have their own information and documentation services that assist outside users as well as member companies. Full names, addresses, and useful details on industrial associations can be found in the *Europa Yearbook* and the *Directory of European Associations,* 2nd ed., Part 1, National Industrial, Trade and Professional Associations, 1976, Part 2, National Learned Scientific and Technical Societies, 1979, I. G. Anderson and G. P. Henderson, eds., CBD Research Ltd., Beckenham, Eng. Other useful sources are the *Encyclopedia of Associations,* 14th ed., N. Yakes and D. Akey, eds., 3 vols., Gale Research Co., Detroit, Mich., 1980; the *National Trade and Professional Associations of the United States and Canada, and Labor Unions,* 15th ed., C. Colgate, Jr. and P. Broida, eds., Columbia Books, Inc., Washington, D.C., 1980; the *World Guide to Trade Associations,* M. Zils, ed., K. G. Saur Publishing Co., New York, 1980.

National Sources.   There is an impressive wealth of economic, statistical, and historical information in the governmental or official publications of each country. Unfortunately for the potential business user, the sheer volume of data makes it difficult to use. Presentation in several languages, in numerous reporting systems, nomenclatures, classifications and codes do not easily lend themselves to useful comparisons. Although international organizations undertake the formidable task of collecting and bringing data together, its usability is still a problem.

In the United States, as in other countries, finding and obtaining governmental publications require considerable expertise. A key to this vast body of information is the *Monthly Catalog of U.S. Government Publications* and the yearly *Directory of U.S. Government Periodicals and Subscriptions Publications.*

The various organs and committees of the United States Congress are sources of fundamental information on the nation and its rules, practices, laws, and policies. The Joint Economic Committee, together with its subcommittees, is responsible for reports and studies on price levels, economic growth, employment, international exchange, foreign economic and trade policies, federal regulations, and economic statistics. Likewise, the various standing committees of the Senate and the House of Representatives deal with all aspects of the nation's economic activity from a legal viewpoint. Select and investigating committees have been responsible for publishing some of the most authoritative documents on the economy.

In Europe, the picture is more complicated because information is published in the national languages and according to the particular rules and practices of the individual countries. Foreign users should seek advice and assistance from national trade associations, chambers of commerce, embassy services, national information centers, and professional information services. Examples of these associations and information centers are the following: ASLIB (Association of Special Libraries and Information Bureaux) in the United Kingdom; the Association Belge de Documentation in Belgium; the French Association des Documentalistes et Bibliothécaires Specialisés; the Deutsche Gesellschaft für Dokumentation in the FRG; the Dansk Teknisk Oplysningstjeneste, which provides assistance to industry in Denmark; the NOBIN (Stichting Nederlands Orgaan, voor de Bevordering vande Informatieverzorging) in the Netherlands. Full names and addresses can be obtained from directories, from the secretariat of the International Federation for Documentation in The Hague, or from the UNESCO Division of General Information Programme in Paris.

The United Kingdom has a long-standing tradition of record keeping. Accordingly, the abundance of source material in the UK is not surprising. British governmental publications are correspondingly numerous, and a useful starting point in locating them is *An Introduction to British Government Publications* by J. G. Olle. Information on official publications concerning the economy appears in the *Brief Guide to Official Publications* published by Her Majesty's Stationery Office (HMSO). These bibliographic tools allow access not only to official publications of direct interest to industry but also to all documents issued by Parliament since 1801.

Until ten years ago the UK Board of Trade issued census publications covering five-year periods. It has now established the Business Statistics Office (BSO) which publishes more current data in its *Business Monitors*. These publications are series of monthly, quarterly, and yearly issues based on information collected by the BSO from industrial firms. These series are described in detail in the *Guide to Short-Term Statistics of Manufacturers' Sales* (Business Monitor PQ 1001) and in the *CSO Guide to Official Statistics* (CSO = Central Statistics Office) published by HMSO. The quarterly publication provides comprehensive and up-to-date sales figures for over 4000 products. In particular, the *Business Monitor Manufacturers' Sales* provides detailed figures on the sales and production of UK establishments that employ 25 or more persons and whose main activity is manufacturing chemicals. It also provides data on exports, imports, employment, production indexes, and wholesale prices. The *Annual Census of Production Monitors* covers practically every sector of the industry and includes data such as total purchases and sales, stocks, work in progress, capital expenditure, employment and wages. The activities of BSO are described in detail in ref. 28. In all these publications, items are classified under the SITC classification system. *The Guide to the Classification for Overseas Trade Statistics*, also published by HMSO, will greatly assist the user. Additional statistical publications of the UK and their issuing agencies appear in Table 2.

In France, as in the United Kingdom, sources of business information are numerous. The Institut National de Statistique et d'Etudes Economiques or INSEE is the organization responsible for collecting, organizing, and publishing economic information. It issues statistical series and short-, middle-, and long-term forecasts. A main publication is the *INSEE Annuaire Statistique de la France,* a statistics yearbook containing tabular and graphical data for the last ten years and covering population, use of natural resources, production of goods including chemicals and allied products, services, prices and wages, and foreign trade and consumption. An outline of the scope and use of the various INSEE publications is given in ref. 29. The INSEE monthlies and other useful statistical publications of France appear in Table 2.

The Benelux countries (Belgium, the Netherlands, and Luxembourg) are unique in the concentration of their main domestic and foreign companies in a small geographical area. These companies, having built large-scale plants for the production of exports, need socio-economic information relating to the countries in which the industries operate and to the importing countries. In this respect, trade statistics are a key instrument. An exhaustive description of economic information in Belgium is given in ref. 30.

The Dutch Ministry of Economic Affairs maintains the Economic Information Service backed by a library of over 80,000 books, many directories and manuals, and

Table 2.   National Statistical Series

Country	Issuing authority	Publication or report	Frequency
United States	Bureau of Census	*Statistical Abstract of the United States*	yearly
		Census of Business	every 5 yr
		Census of Manufacturers	every 5 yr
		Census of Production	every 5 yr
		Census of Transportation	every 5 yr
		Census of Government	every 5 yr
	Office of Business Economics	*Survey of Current Business*	monthly
	U.S. Department of Commerce, Industry and Trade Administration	*International Economic Indicators*	quarterly
	Bureau of International Commerce	*Overseas Business Reports*	irregular
		Market for U.S. Products	irregular
		International Commerce	irregular
	Department of Commerce	*Business Service Checklist*	weekly
		Monthly Catalog of U.S. Government Publications	monthly
	Department of Labor	*Monthly Labor Review*	monthly
	Bureau of Labor Statistics	*BLS Reports*	irregular
		BLS Bulletins	irregular
		Manpower Report	yearly
	Congressional Information Service	*American Statistics Index: A Comprehensive Guide and Index to the Statistical Publications of the U.S. Government*	yearly
	U.S. Government Publications Office	*Business Conditions Digest*	monthly
	U.S. Council of Economic Advisors	*Economic Indicators*	quarterly
	Board of Governors Federal Reserve System	*Industrial Production*	monthly
		Federal Reserve Bulletin	monthly
United Kingdom	Business Statistics Office (BSO)	*Business Monitors*	monthly, quarterly, yearly
	Central Statistical Office (CSO)	*Monthly Digest of Statistics*	monthly
		Annual Abstracts of Statistics	yearly
	Department of Trade and Industry	*Overseas Trade Statistics of the United Kingdom*	monthly, yearly, every 2 yr
		Trade and Industry	weekly
		British Business	weekly
		Statistics of Trade through United Kingdom Parts	quarterly
France	Institut National de Statistique et d'Etudes Economiques (INSEE)	*Economie & Statistiques*	monthly
		Tendances de la Conjoncture-Graphiques Mensuels	monthly
	Ministère de l'Industrie (Ministry of Industry)	*Annuaire de Statistiques Industrielles*	yearly
		Bulletin Mensuel de Statistiques Industrielles	monthly

Table 2 (*continued*)

Country	Issuing authority	Publication or report	Frequency
	Direction Générale des Douanes et Droits Indirects	*Statistiques du Commerce Extérieur*	yearly
Belgium	Institut National de Statistiques (INS)	*Annuaire Statistique de la Belgique*	yearly
		Statistiques Industrielles	monthly
		Communiqué Hebdomadaire–Weekbericht	weekly
		Statistique Fiscale des Revenus	yearly
	Banque Nationale de Belgique (National Bank of Belgium)	*Statistiques Economiques de Belgique*	monthly
	Ministère des Affaires Economiques (Ministry of Economy)	*Bulletin Mensuel du Commerce Extérieur*	monthly
The Netherlands	Central Bureau of Statistics (CBS)	*Statistical Yearbook (Jaarcijfers)*	yearly
		Statistisch Zakboek	yearly
		Maandstatistiek van de Industrie	monthly
		Maandschrift	monthly
		Statistisch Bulletin	irregular
		Chemische Industrie-Produktiestatistieken	yearly
		Rubberverwerkende Industrie-Produktiestatistieken	irregular
		Maandstatistiek van Buitenlandse Handel	monthly
Federal Republic of Germany	Statistisches Bundesamt	*Statistisches Jahrbuch für die Bundesrepublik Deutschland*	yearly
		Produzierendes Gewerbe	quarterly
		Aussenhandel nach Waren und Ländern	monthly
	IFO-Institut für Wirtschaftsforschung	*Wirtschafts-Konjunktur*	monthly
		IFO Schnelldienst	weekly
	Federal Minister of Economics	*The Economic Situation in the Federal Republic of Germany*	monthly
	Verband der Chemischen Industrie e.V.	*Jahresbericht*	yearly
		Chemiewirtschaft in Zahlen	yearly
Italy	Instituto Centrale di Statistica (ISTAT)	*Annuario Statistico Italiano*	yearly
		Bollettino Mensile di Statistica	monthly
		Annuario di Statistiche Industriali	yearly
		Indicatori Mensili	monthly
		Notiziario ISTAT	monthly
		Statistica Mensile del Commercio con l'Estero	monthly
	Instituto Nazionale per lo Studio della Congiuntura (National Institute for Economic Research)	*Congiuntura Italiana*	monthly
		Quaderni Analitici	biweekly
Norway	Nordisk Råd (Nordic Council)	*Yearbook of Nordic Statistics*	yearly

Table 2 (*continued*)

Country	Issuing authority	Publication or report	Frequency
		Statistik Årbok	yearly
		Utenrikshandel	yearly
Denmark		*Statistik Årbog*	yearly
		Varestatistik	yearly
		Danmarks Vareindførsel og-udførsel	yearly
Finland		*Suomen Tilastollinen Vuosikirja*	yearly
		Ulkomaankauppa	yearly
Sweden		*Statistik Årsbok*	yearly
		Utrikeshandel	yearly
Switzerland	Eidgenössisches Statistisches Amt (Federal Bureau of Statistics)	*Statistisches Jahrbuch der Schweiz-Annuaire Statistique de la Suisse*	yearly
	Eidgenössische Oberzoll-direktion (Customs)	*Jahresstatistik des Aussenhandels der Schweiz*	yearly
Austria	Österreichisches Statistisches Zentralamt	*Statistisches Handbuch*	yearly
		Industriestatistik	yearly
		Der Aussenhandel Österreichs	yearly
Spain	Instituto Nacional de Estadística	*Anuario Estadístico de España*	yearly
		Estadísticas de Producción Industrial	monthly, yearly
	Comisión Asesora y de Estudios Técnicos de la Industria Química Española	*La Industria Química en España*	yearly
		Situación y Perspectivas de la Industria Química Española	yearly
		Estadística del Comercio Exterior de España	yearly
Portugal	Instituto Nacional de Estatística	*Anuário Estatístico*	yearly
		Estatísticas Industriais	yearly
		Estatísticas do Comércio Externo	yearly
Greece	National Statistical Service of Greece	*Statistical Yearbook of Greece*	yearly
USSR	Council for Mutual Economic Assistance	*Statistical Yearbook*	yearly

over 3000 periodicals. The Service publishes *Economic Titles,* a semimonthly bulletin on the economy of the Netherlands and of other countries and regions. The Service carries out literature searches and provides rapid reference service on request. This main source of economic information in the Netherlands and also those in Belgium are described in ref. 31.

In the Federal Republic of Germany, the Statistisches Bundesamt in Wiesbaden is the central organization that by law issues data on a national level. Different ministries, in their regional and local offices, also collect and publish statistics. As a result there are hundreds of German-language publications available to the public. The main source publications are named in Table 2.

Foreign trade statistics for imports and exports are more readily obtainable from authorities that have computerized their operations. The user may either subscribe to the monthly update tapes and process these in-house or obtain monthly computer printouts that provide data on the commodities of interest. The names and addresses of some of the major foreign trade-data sources are the following: Institut National de Statistiques, Rue de Louvain 44, 1000-Bruxelles, Belgium; Direction Générale des Douanes, Centre de Renseignements Statistiques, 192 Rue St. Honoré, 75056 Paris, France; Statistisches Bundesamt, Gustav Stresemann—Ring, 11, Postfach 5528, 6200 Wiesbaden, FRG; Centraal Bureau voor de Statistiek, Statistieken van de Buitenlandse Handel, Prinses Beatrix laan 428, Voorburg, The Netherlands; Instituto Centrale di Statistica, Direzione Generale dei Servici Tecnici, Via Cesare Balbo 16, 00 100 Rome, Italy; and H. M. Customs and Excise Statistical Office, Portcullis House, 27 Victoria Avenue, Southand-on-Sea SS2 6AL, UK.

Information on finance, taxes, and monetary economics may cover broad as well as specific subject matter; public finance encompasses governmental and industrial activities, national issues such as nationalization or the financing of local projects, budgeting, fund raising, taxation of corporate or private income, the restriction on movement of funds abroad, currency stability and rates of exchange, reevaluation and devaluation, and inflation. Numerous specialized publications provide rapid and reliable information on these matters.

The main banks in many countries are known for their publications and some for their outstanding information services. They publish bulletins on the financial and economic situation, domestic and foreign, and reviews on the outlook and trends of various industrial sectors. The *Bankers' Almanac and Yearbook,* Thomas Skinner and Co., Ltd., London, covers banks on an international scale. Some noteworthy bank publications are:

Belgium: Banque Nationale—*Bulletin;* Banque Bruxelles-Lambert—*Bulletin Commercial, Bulletin Financiere, Reports from Brussels;* Societe Générale de Banque—*Bulletin de la Société Générale de Banque, Conjoncture Boursière;* Kredietbank—*Weekly Bulletin;* Banque de Paris et des Pays-Bas—*Parisbas Belgique Présente.*

France: Banque de France—*Enquête Mensuelle de la Conjoncture, Bulletin Trimestriel de la Banque de France, Cahiers Economiques et Monetaires.*

Germany: Deutsches Bundesbank—*Monthly Report of the Deutsches Bundesbank, Statistische Beihefte zu den Monatsberichten der Deutschen Bundesbank.*

Italy: Banca d'Italia—*Bolletino Statistico Trimestrale, Contributi alla Ricerca Economica;* Banca Nazionale del Lavoro—*Italian Trends;* Banco di Roma—*Review of the Economic Conditions in Italy.*

Spain: Banco de España—*Boletin Estadístico, Informe Anual;* Banco de Bilbao—*Servex.*

The Netherlands: Amsterdam-Rotterdam Bank—*Amro Economisch Bulletin, Dutch Economy in Figures, Netherlands Economic Report.*

United Kingdom: Bank of England—*Quarterly Review, Bank of England Quarterly Bulletin.*

United States: Citibank—*Monthly Economic Letter;* Morgan Guaranty Trust Co.—*World Financial Markets.*

Denmark: Danmarks Nationalbank—*Danmarks Nationalbank Report and Accounts.*

In addition to the publications from official bodies and specialized organizations, newspapers and periodicals are primary sources of current information. Press agencies and national information offices of each country may provide guidance in the selection of titles most appropriate to the user's requirements. Publications pertinent to chemical industries are listed in Table 3. A minimal reading list is: *The Financial Times; The Journal of Commerce* (U.S. Edition); *The Journal of Commerce* (European edition); *The Wall Street Journal; The Frankfurter Allgemeine Zeitung; Il Corriere della Serra; Les Echos; Le Monde; Nachrichten für Aussenhandel; L'Echo de la Bourse-AGEFI;* and *Neue Zürcher Zeitung.* The following references are most useful for further details and source material.

DAFSA—Centre Français du Commerce Extérieur, *L'Information Economique et Financière en Europe—Répertoire des Sources*, DAFSA (S.A. de Documentation et d'Analyses Financieres), Paris, Fr.
Eleonore de Dampierre et François Charpentier, *Les Sources d'Information Economique et Financière: Etats-Unis, Royaume-Uni, France*, Center for Business Information, Paris, Fr.
J. Fletcher, ed., *The Use of Economics Literature*, Butterworths Pub., Inc., Woburn, Mass., 1971; a most illuminating and instructive book.
Published Data on European Industrial Markets, Industrial Aids, Ltd., London, Eng. 1975; an outstanding directory of source material and its contents.
L. Daniells, *Business Information Sources*, Center for Business Information, University of California Press, Berkeley, Calif., 1976.
J. M. Harvey, *Statistics—Europe: Sources for Social, Economic and Market Research*, 3rd ed., CBD Research Ltd., Beckenham, Eng., 1976, Gale Research Company, Detroit, Mich.
J. M. Harvey, *Statistics—Europe: Guide for the Market Research to 34 Countries of Europe*, 3rd ed., International Publications Service, New York, 1976.
B. Lawrence, "Preliminary Project Evaluation—Any Technologist Can Do It. Guide to Chemical Business Information Sources," *Chemtech* **5,** 678–681 (1975).
K. D. Vernon, ed., *Use of Management and Business Literature*, Butterworths Pub., Inc., Woburn, Mass., 1976.
N. Roberts, *Use of Social Sciences Literature*, (*Information Sources in Sciences and Technology Series*), Butterworths Pub., Inc., Woburn, Mass., 1976.
F. C. Pieper, *SISCIS, Subject Index to Sources of Comparative International Statistics*, CBD Research, Ltd., Beckenham, Eng., 1978; very useful.

The British Overseas Trade Board publishes a series of pamphlets entitled *Hints to Businessmen (Country Name)* that contains general information on the country that business visitors may need to know, economic factors, import and exchange control regulations, methods of doing business, and a list of governmental and commercial organizations.

Sources of European Economic Information, 2nd ed., compiled by Cambridge Information and Research Services, Ltd., Gower Press, distributed by Unipub, New York, 1978, is an easy-to-use reference to source publications.

Market and Product Information. Many resources are available for obtaining information on markets and products. In-house market researchers should seek the assistance of professional marketing consultants when the information required cannot be readily prepared in-house or when these consultants have the needed expertise in fields outside the concerns of the user. Much information is also obtainable from official agencies described above, particularly industrial associations, chambers of commerce, and from periodicals, trade magazines, and the press. Specialized and specific trade associations are important sources of information. Among the most important of these are the Association of British Pharmaceutical Industry, the Paint Maker Association of Great Britain, the British Agrochemicals Association, the Fertilizer Manufacturers' Association, the Hydrocarbon Solvents Association, the British Compressed Gases Association, the British Colour Makers Association, and the National Sulfuric Acid Manufacturers' Association (see Market and marketing research).

The individual countries' foreign-trade offices are a source of market information abroad. Main offices include Office Belge du Commerce Extérieur (Belgium), Centre Français du Commerce Extérieur (France), Economische Voorlichtingsdienst (the Netherlands), Bundesstelle für Aussenhandels Information (FRG), and Office Suisse d'Expansion Commerciale (Switzerland). These offices collect international information on economies, markets, and products from periodicals, dailies, and original documents. This information is then indexed, stored, and disseminated as printed publications or microfiche. Their libraries store large collections of foreign trade statistics, catalogs, directories, and reference material and well organized market files, business opportunities files, and foreign importers files. These offices are also a source of business and marketing information on countries in the Middle East, Africa, and particularly countries with centrally planned economies. In this respect, the daily, *Nachrichten für Aussenhandel,* published by the German Bundesstelle für Aussenhandels Information, which provides statistical and national supplements, is outstanding. The foreign-trade offices also have documentation and information centers from which assistance may be obtained, the addresses and scope of services of which are provided in the DAFSA *L'Information Economique et Financière en Europe— Répertoire des Sources.*

Chambers of commerce, special agencies, and offices established in most countries are another source of market information. Names and activities of these organizations appear in publications such as the *Encyclopedia of Business Information Sources,* 3rd ed., P. Wasserman and co-eds., Gale Research Company, Detroit, Mich., 1977; *Marketing and Management: A World Register of Organizations,* I. G. Anderson, CBD Research, Ltd., London, Eng., 1969; International Publications Service, New York, 1969; *Published Data on European Industrial Markets,* 3rd ed., Industrial Aids, Ltd., London, Eng., 1974.

Marketing research firms and consultants are unique sources of market information. The *Directory of U.S. and Canadian Marketing Surveys and Services,* K. Di Cioccio, and V. Kollonitsch, eds., Charles H. Kline & Co., Inc., Fairfield, N.J., 1976, lists about 1500 multiclient marketing reports and describes the scope of services of 125 consulting firms. This publication is supplemented by the *International Directory of Published Market Research* compiled by the British Overseas Trade Board, Ltd., London, Eng., 1976. It has a detailed alphabetical product index; studies and publications are entered under the British Standard Industrial Classification scheme so that they are grouped by subject. Products and services covered by these studies are briefly described and include geographical coverage, date, source, and price.

The following directories, guides, handbooks, business research reports, product studies, and reports series are useful for markets and products information.

D. Degen and T. E. Miller, eds., *Findex: the Directory of Market Research Reports, Studies and Surveys, 1979,* Information Clearing House, Inc., New York; reference guide to more than 2500 commercially available market and business research reports produced internationally.

Management Information Guide Series, Gale Research Co., Detroit, Mich.; deals with specialized areas of business and industry.

L. J. Wheeler, ed., *International Business and Foreign Trade Information Sources,* Gale Research Co., Detroit, Mich., 1968.

J. V. Kopycinski, *Textile Industry Information Sources,* Gale Research Co., Detroit, Mich., 1964.

Food and Beverage Industries: A Bibliography and Guidebook, A. C. Vara, Gale Research, Detroit, Mich., 1970.

T. Landau, ed., *European Directory of Market Research Surveys,* Gower Press, distributed by Teakfield, Ltd., Hampshire, Eng.; Unipub, New York, 1976; references about 1500 market research reports published since 1972.

Table 3.   Publications Pertinent to Chemical Industries

Country	Frequency	Publication
Belgium	dailies	*L'Echo de la Bourse-AGEFI*
		Europe
		Financieel Economische de Tijd
		Le Lloyd Anversois
		Official Journal of the EEC
	weeklies	*Banque Bruxelles Lambert*
		Bulletin Commercial
		Bulletin Financiere
		European Report
		Kredietbank Bulletin
		Le Marché
	fortnightlies	*Fédération des Entreprises de Belgique—Bulletin*
	monthlies	*Aperçu de l'Evolution Economique*
		Banque Bruxelles Lambert—Bulletin de Conjoncture
		Banque Nationale de Belgique—Bulletin
		Belgian Business
		Belgian Trade Review
		Commerce in Belgium
		Distribution d'Aujourd'hui
	quarterlies	*Annales des Sciences Economiques Appliquées*
		Cahiers Economiques de Bruxelles
		Recherches Economiques de Louvain
	annual	*Banque Bruxelles Lambert—Rapport Annuel*
	irregular	*Economic Situation in the Community*
France	dailies	*Les Echos*
		Le Monde
	weeklies	*M.O.C.I. Moniteur du Commerce International*
		L'Usine Nouvelles
	fortnightlies	*Arab Oil and Gas*
		Chimie Actualités
		L'Observateur de l'OECD
		OECD Observer
		Parfums, Cosmétiques et Arômes
		Revue Générale Africaine
	monthlies	*La Corderie Française*
		Emballages
		L'Expansion
		Industrie des Plastiques et Elastoméres
		L'Industrie Textile
		Information Chimie
		Information d'Outre-Mer
		Main Economic Indicators
		Le Nouvel Economiste
		L'Officiel des Plastiques et du Caoutchouc
		Plastiques Modernes et Elastoméres
	quarterlies	*European Business*
	irregular	*OECD Economic Outlook*
		Revue du Marché Commun
FRG	dailies	*Frankfurter Allgemeine Zeitung*
		Handelsblatt
		Nachrichten für Aussenhandel
	fortnightlies	*Bundesgesundheitsblatt*
		Europa Chemie
	monthlies	*Chemie Ingenieur Technik*

Table 3 (*continued*)

Country	Frequency	Publication
		Chemische Industrie
		Chemie Fasern—Textile Industrie
		Erdöl und Kohle
		Farbe und Lack
		German International
		Die Gummibereifung
		Kautschuk und Gummi-Kunstsoffe
		Kunststoffberater-Rundschau
		Kunststoffe—German Plastics
		Melliand Textilbericht
		O E L
		Plastverarbeiter
		Wirtschaft und Statistik
Israel	quarterlies	*Reviews on Environmental Health*
Italy	dailies	*Corriere della Sera*
		24 Ore
	fortnightlies	*Industria Chimica*
		L'Industria Italiana del Plastici
	monthlies	*Chimica e l'Industria*
		Espansione
		L'Industria della Gomma
		L'Industria della Vernice
		Materie Plastiche
Japan	weeklies	*Japan Chemical Week*
	fortnightlies	*Japan Plastics*
	monthlies	*Chemical Economy and Engineering Review*
		Japan Textile News
		Plastics Industry News (Japan)
The Netherlands	weeklies	*Chemisch Weekblad*
		Chempress
		Wereldmarkt
	monthlies	*Plastica*
	quarterlies	*Common Market Law Review*
Spain	weeklies	*Actualidad Económica*
		El Economista
	monthlies	*I.Q. (Industria Química)*
Sweden	weeklies	*Veckans Affarer*
Switzerland	dailies	*Neue Zürcher Zeitung*
	weeklies	*Chemische Rundschau*
United Kingdom	dailies	*Financial Times*
		Journal of Commerce
		Lloyd's List
		Lloyd's Shipping Index
	weeklies	*Chemical Age*
		Chemist and Druggist
		The Economist
		European Chemical News
		Fairplay
		IMS Pharmaceutical Market Letter
		Investors Chronicle
		Marine Week
		Marketing in Europe
		Metal Bulletin
		Middle East Economic Digest
		New Scientist

Table 3 (*continued*)

Country	Frequency	Publication
		Pharmaceutical Journal
		Plastics and Rubber Weekly
		Scrip
		Shoe and Leather News
	fortnightlies	*Africa Confidential*
		Agra Europe
		Chemical Insight
		Eurolaw Commercial Intelligence
		International Dyer
		International Flavours and Food Additives
		International Pest Control
		Nitrogen
		Petroleum Times
		Phosphorous and Potassium
		Polymers, Paint, and Colour Journal
		Sulphur
		Wool Record and Textile World
	monthlies	*British Plastics and Rubber*
		Confectionary Production
		Continental Paint and Resin News
		Cordage, Canvas and Jute
		Dairy Industries International
		European Industrial Relations Reviews
		European Plastics News
		European Rubber Journal
		Fertilizer International
		Food Manufacture
		Food Processing Industries
		Furniture Manufacturer
		Information—CEE
		Journal of Flour and Feed Milling
		Manufacturing Chemist and Aerosol News
		New African
		Packaging News
		Packaging Review
		Petroleum Economist
		Pigment and Resin Technology
		Reinforced Plastics
		Soap, Perfumery, and Cosmetics
		Soft Drinks Trade Journal
		Textile Month
		Textile Institute and Industry
	quarterlies	*British Ink Maker*
		Futures
		Rubber Trends
	irregular	*Plastics Today (ICI)*
United States	dailies	*Wall Street Journal*
		Journal of Commerce
	weeklies	*Business Europe*
		Business International
		Business Week
		Chemical and Engineering News
		Chemical Marketing Reporter
		Chemical Week
		Feedstuffs

Table 3 (*continued*)

Country	Frequency	Publication
		The Guardian
		Oil and Gas Journal
		Petrochemical News
	fortnightlies	*Business Eastern Europe*
		Chemical Engineering
		Fortune
		Naval Stores Review
	monthlies	*Adhesives Age*
		Chemical Engineering Progress
		Chemtech
		Cotton
		Cosmetics and Toiletries
		Drug and Cosmetic Industry
		Farm Chemicals
		Food Technology
		Household and Personal Products Industry
		Hydrocarbon Processing
		Insulation
		Modern Packaging
		Modern Plastics International
		Modern Textiles
		National Safety News
		Near East Business
		Pulp and Paper International
		Rubber World
		Soap, Cosmetics, and Chemical Specialties
		Textile Chemist and Colorist
		Textile Organon

Process Evaluation/Research Planning (PERP reports), Chem Systems, Inc.; comprehensive studies, each covering a type of chemical product and the markets by national regions; updated by the *PERP Quarterly Reports.*

Orthoxylene–Phthalic Anhydride Annual World Survey, the *para-Xylene–DMT/PTA Annual World Survey,* and the *World Naphtha Survey Capacity Tables–Petrochemical Plants* are an interesting series of product studies offered by the HYPLAN Consulting Group.

Eastern Europe—Its Impact on the West European Chemical Industry, 1978, is a unique HYPLAN report that contains information on the production of base petrochemicals, first and second order petrochemical derivatives, and the main product in Bulgaria, Czechoslovakia, the German Democratic Republic, Hungary, Poland, Romania, the USSR, and Yugoslavia.

Petrochemical Units in Western Europe and *Petrochemical Units in the OPEC and OAPEC Countries,* Institute Français du Petrole, Paris; updated occasionally.

Industria Petrolchimica Europea (IPE), Parpinelli-Tecnon, Italy.

The *Chemeurop–EEC* report series, SEMA (Metra International), provides statistical data for about 130 chemicals and derivatives in terms of production, producers, location, current and projected capacities, and total consumption breakdown by main uses.

SRI International offers diverse services to its subscribers from a wide range of industries. SRI's Business Information Program provides in-depth studies and forecasts covering business, economic, political, social, and technological topics on a regional or international basis. SRI developed a number of information resources specifically for the chemical, petroleum, and engineering sectors. These include the following:

Process Economics Program (PEP) report series, SRI International; evaluates the processes for the production of chemicals and petroleum products.

Chemical Economics Handbook (CEH), SRI International; provides data on production, sales, exports, imports, consumption, stocks and shipments of raw materials, primary and intermediate chemicals, or end products such as fertilizers or polymers.

The *World Hydrocarbons Program*, 1979, developed by SRI International; series of reports on approximately 60 main chemicals of interest to the international petrochemical industry.

Directory of Chemical Producers—USA and *Directory of Chemical Producers—Western Europe*, 1979, SRI International provide information on companies and products, list chemicals by manufacturer and plants by location.

The National Economic Development Office (NEDO) in the UK has issued several reviews on industrial sectors and monographs on manufacturing industries in the UK and in EEC countries. Some of the titles are listed in T. Landau's *European Directory of Market Research Surveys*.

Principal Sources of Marketing Information, a booklet by C. Hull of the *Times (London)* Information and Marketing Intelligence Unit, Oct. 1975, brings together the principal sources of information on markets for several industries.

Sources and Production Economics of Chemical Products 1979, compiled and edited by *Chemical Engineering*, McGraw-Hill Publications Company, New York, 1979.

European Chemical Industry Handbook, Hedderwick Stirling Grumbar, UK, 1979; a compilation of data on production, trade, and finance of the chemical industry.

Handbooks and annual installments of these and other series rapidly become dated. Therefore, specialized information services must continuously capture, index, disseminate, and store current business, economic, marketing, and industrial information from periodicals such as those listed in Table 2. Although this list is not exhaustive, it does show the most useful sources of business information. The daily publications, *Platt's Oilgram Price Report,* which indicates the prices of petroleum products internationally, and *Petrochemical Scan,* which covers the base petrochemicals and is distributed via telex networks, both include timely informational briefs on major events that influence petroleum products.

Company Information. Although available information on companies is diverse, the most common needs are for company names, addresses, and products. Information requested next is usually for a company's performance including data such as turnover, sales, profit before and after taxes, capital expenditures, number of employees, ownership, number of subsidiaries, and facilities. Intercompany comparisons may require that the financial results be analyzed in more detail thus requiring the examination of balance sheets and historical data to reveal trends.

Company names, addresses, ownership, and products appear in directories and handbooks that have international or national coverage. Some of these are listed below.

Who Owns Whom, Dun & Bradstreet, Ltd., London, Eng. lists the parent and associated companies in a series of volumes covering the world.

The Worldwide Chemical Directory, prepared by ECN Chemical Data Services, IPC Business Press Ltd., New York, 1977 provides the address, telex, and the telephone numbers of chemical companies and a summary of their activities.

Chem Sources—U.S.A., 21st ed., Directories Publishing Co., Inc., Flemington, N.J., 1980, and *Chem Sources—Europe,* Chemical Sources Europe, Mountain Lakes, N.J., 1980 provide information on manufacturers or suppliers of chemical products.

OPD Chemical Buyers Directory, 67th ed., Schnell Publishing Co., Inc., New York, 1979 (appearing once a year as part of an annual subscription to *Chemical Marketing Reporter*) provides information on chemicals, their suppliers, shipping, and storage.

1980 Chemical Week Buyers' Guide Issue, McGraw-Hill, Inc., New York (appearing once a year as part of an annual subscription to *Chemical Week*) gives information on chemicals, new materials, specialties, suppliers, trade names, company addresses, and directors' names.

Directory of Chemical Producers—United States and *Directory of Chemical Producers—Western Europe,* SRI International, Menlo Park, Calif., 1979; provide company names, addresses, ownerships, subsidiaries, plants, and products and list the chemicals by manufacturing company and plants by location.

KOMPASS, Kompass Register, Ltd., Croydon, Eng.; series of directories for individual countries, especially those in Europe; products and company indexes, supplier names, addresses, and business activities, employees, and a trade name index are provided.

Directory of West European Chemical Producers, 1977–1978 ed., Chemical Information Services, Ltd., Oceanside, N.Y. provides information for about 28,000 products and their suppliers.

European Chemical Buyers' Guide, IPC Business Press, Ltd., New York, 1979 provides product, company and supplier, and trade name indexes.

More detailed information on the ownership, shareholders, capital, industrial operations, and finances appears in directories such as the *Financial Times International Business and Companies Year Book: 1978/79,* B. E. Donovan, ed., The Financial Times, Ltd., London, Eng., 1978–1979. This book quotes the top companies worldwide, their addresses, subsidiaries, ownership, capital results, main products, and gives a brief overview of the individual countries' economies.

Comprehensive information on companies appears in the following sources:

R. P. Hanson, ed., *Moody's Industrial Manual,* Moody's Investors Service, Inc., New York, 1979 covers all United States industrial sectors.

J. Love, ed., *Jane's Major Companies of Europe 1979–80,* Macdonald and Jane's Publishers, Ltd., London, Eng.; Franklin Watts, Inc., New York, 1979.

A number of official agencies and specialized services provide information on corporations and companies for investment, financial analyses, and accounting and legal matters. Because the information they deal with is required by law, these agencies are reliable sources of these particular types of business data. In the United States this type of information is filed with the Securities and Exchange Commission (SEC). The SEC issues several comprehensive reports, all of which would not be possible to cover in detail here. Also worth mentioning are the 10-K annual business and financial reports. These identify the main products and services of a company and summarize its operations in detail for the last five fiscal years. The 10-Q quarterly financial reports filed by most companies provide current information on their financial position during the year. The 8-K reports inform shareholders of important events or changes such as acquisition of new assets, increase or decrease in the amount of securities outstanding, and the like.

In the United Kingdom, companies registered in England and Wales are required to file certain information with the Companies Registration Office and Companies House; those registered in Scotland must file with the Register of Companies. In France, similar information can be obtained from the official publication, *Bulletin des Annonces Légales Obligatoires* (BALO); in Germany, Verlag Hoppenstedt publishes economic and financial information on companies. Names and addresses of agencies and services that publish or file company information appear in DAFSA's *L'Information Economique et Financière en Europe, Les Sources d'Information Economique et Financière: Etats-Unis, Royaume-Uni, France,* E. de Dampierre and F. Charpentier, Center for Business Information, and *Published Data on European Industrial Markets,* Industrial Aids, Ltd.

Two specialized services worthy of note are the DAFSA–Hoppenstedt *Information Internationales* and the British Extel Statistical Service, Ltd. The former provides information on large companies in loose-leaf form printed in French, German, and English. The latter publishes information on cards covering mostly British companies but including also some major European companies.

Of course, the daily press and periodicals such as those listed in Table 2 are sources for recent news. Secondary services also provide information on company activity internationally. The best known for the chemical industrial sector are Chemical Abstracts Service's *Chemical Industry Notes* (*CIN*), the American Petroleum Institute's *Petroleum/Energy Business News Index,* and Predicasts' *Prompt* and *F&S Index.*

Computerized Information Sources. Computerized information files, accessible on line from remote terminals, constitute a distinct body of information that has grown rapidly in recent years. Searching requires expertise, particularly in the area of business information, because of the diversity of available files. Many are bibliographic, but some technical-economic files, resembling the numerical or nonbibliographic files discussed in the previous section of this article, provide business information directly. Some systems offer manipulative and computational capabilities that allow the searcher to store data, enter data, merge or modify series of data, manipulate them, edit results on line, and print these in report format. An advantage of these systems is that they use natural-language commands; the searcher needs no programming training or knowledge of programming language. But because a large number of data bases are available, searchers should develop a familiarity with those files that most often meet their needs, although other files, upon exploration, may unexpectedly provide additional information. Searchers must be on the alert continuously for new files and services and must be able to assess their value to the organizations they serve.

Table 4 outlines some business-information services and data bases currently available. The data bases listed are those important to chemical business, management, and corporate policy. The listing excludes those data bases that are primarily scientific and technical in nature.

Nonbibliographic Business Data Bases. Data Resources, Inc. (DRI), in Boston, Mass., makes available information on the economies of several countries from data bases that contain time series from United States, foreign governmental and nongovernmental sources and organizations. DRI's services allow the user to relate external economic developments to the corporate planning of manufacturing companies in terms of future sales, cash flow, capital budgeting and investment, inventories, production, pricing, business cycles, or the implications of governmental decisions. Furthermore, DRI has developed economic models and forecasts for individual regions, ie, the United States, Canada, the European Economic Communities, and Japan.

DRI's source material includes economic, financial, and demographic series on the United States; the complete details of consumer, wholesale, and industrial price indexes as compiled by the Bureau of Labor Statistics; a complete set of OECD series pertaining to economic indicators, national income accounts, trade series, industrial production inclusive of chemicals, petroleum and coal products; and international financial statistics from the International Monetary Fund. DRI's data bases can be searched in a variety of ways, data can be manipulated on line, and several report-writing capabilities are available. Figure 4 illustrates a feature that permits the user to draw a three-month moving-average line from dispersed monthly figures.

Chase Econometrics provides access to several data bases and to econometric models with full simulation capability. These models allow the user to link the economies of the following countries: the United States, FRG, Belgium, France, Italy, the Netherlands, the United Kingdom, Spain, Sweden, Japan, Canada, Mexico, and Brazil. The system also comprises historical and forecast data bases. The historical data bases include detailed financial series and wholesale and consumer price series. The forecast data bases contain Chase Manhattan's estimates as derived from their models.

Chase Manhattan also has a computerized company data-base service, called EXSTAT from the Extel Statistical Services, Ltd., of London, which provides information on the United Kingdom's, continental European, and Australian companies.

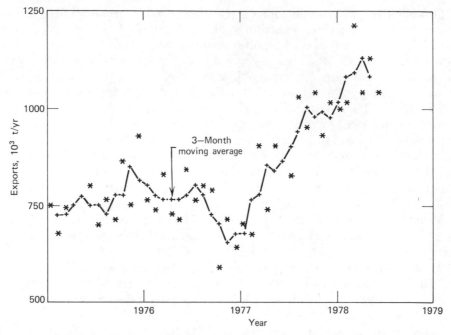

Figure 4. Exports of five main plastics. Courtesy of Data Resource, Inc.

The information comes from the same source as the Extel Cards service but can be analyzed interactively on line through Interactive Data Corporation's (IDC) time-sharing network.

Automatic Data Pioneering, Inc. (ADP), through ADP Network Services International and Information Services, offers files including the EXSTAT data base from Extel, COMPUSTAT, which provides financial information on the United States, COPPER DEVELOPMENT DATA BASE, McGraw-Hill's METALS WEEK, and the IPC CHEMICAL DATA BASE, which covers plant and international product information on chemicals.

Parpinelli-Tecnon, in Milan, Italy, computerized their petrochemicals file and made it accessible on line under the name EPICS (European Petrochemical Industry Computerized System). The statistics segment of the file provides the complete balance for each product for selected countries or regions from 1965 to the current year, for the next five years (forecasts), and an output eighth year. The plants section provides information on each selected plant or product in terms of producing company, location, feedstock(s), plant status, start-up date, license, contractor, and capacity for the same time period as the statistics section. Program features allow users to retrieve information on an entire company group (ie, the mother company with all its subsidiaries) or on all plants in the same location or region that manufacture a particular product. The program also allows the user to analyze supply-and-demand balances for each product by region, country, location, or individual company. In carrying out this analysis, users can manipulate the retrieved data by adding or replacing some of it with their own figures. An example of data provided by the EPICS data base is shown in Table 5.

Stanford Research Institute's (SRI) WORLD PETROCHEMICAL DATA BASE

Table 4. Business Data Bases

Data base name	Supplier	Coverage
ABI–INFORM	Data Courier, Inc.	business management and administration; 1971–present
ACCOUNTANTS' INDEX	American Institute of Certified Public Accountants	international accounting and finance; 1974–present
AFO (Analyse Financière sur Ordinateur)	CISI, France	financial data for quoted companies in France
APTIC	U.S. Environmental Protection Agency	air pollution in the broad sense, including social, political, legal, administrative, and technical aspects; 1966–1978
ASI (American Statistics Index)	Congressional Information Service, Inc.	abstracts and indexes of all U.S. federal statistical publications; 1973–present
BIPA (Banque d'Information Politique et d'Actualité)	La Documentation Française and Télesystèmes, France	several files of official French publications on a variety of economic and political issues
CBPI	Information Access of Toronto, Canada	Canadian business periodicals; 1975–present
CEE (Archivio della Giurisprudenza della Corte Di Giustizia della Comunita Europee)	Centro Elettronko di Documentazione della Corte Suprema di Cassazione, Italy	file on jurisprudence containing all the digests of the decisions of the EEC's Court of Justice
CELEX	Legal Service of the Commission of the European Economic Communities	legislative and legal information for member countries; anticipated in 1980; 1977–present
CIN (Chemical Industry Notes)	Chemical Abstracts Service	chemical processing industries; 1974–present;
CIS INDEX	Congressional Information Service, Inc.	current, comprehensive index to the entire U.S.; congressional working papers; 1970–present
COMPUSTAT (Files)	Standard and Poor's Compustat Services, Inc.	several nonbibliographic files containing financial information on the U.S. and Canada
COPPER DEVELOPMENT ASSOCIATION	Copper Development Association	nonbibliographic information on metals
CRECORD	Capitol Services, Inc.	current coverage of the *Congressional Record*—all proceedings of the U.S. Congress; 1976–present
CRONOS (Chronological Series)	Statistical Office of the Commission of the European Economics Communities	time series of the member countries of the European Economic Communities; may become a public file in 1980
DAFSA–LES LIAISONS FINANCIERES	Société Pour d'Informatique, France	information on corporate relationships between companies in France
DISCLOSURE	Disclosure Incorporated	securities and exchange Commission filings; 1977–present
DRI CAPSULE DATA BASE	Data Resources, Inc.	business and economics data
DRI CENTRAL DATA BANK	Data Resources, Inc.	macroeconomics, primarily U.S. data
ECONOMICS ABSTRACTS INTERNATIONAL	Learned Information, Ltd., UK	international coverage of markets, industries, country-specific economics and research in management and economics; 1974–present

51

Table 4 (*continued*)

Data base name	Supplier	Coverage
EIS INDUSTRIAL PLANTS	Economic Information Systems, Inc.	current nonbibliographic information on establishments representing over 90% of U.S. industrial activity; replaced three times per year
EIS NONMANUFACTURING ESTABLISHMENTS	Economic Information Systems, Inc.	current data on U.S. nonmanufacturing establishments employing 20 people or more; replaced three times per year
ENCYCLOPEDIA OF ASSOCIATIONS	Gale Research Company	detailed information on several thousand trade associations, professional societies, etc; current year
ENERGYLINE	Environment Information Center, Inc.	socio-economic, governmental policy and planning, current affairs, and scientific–technical aspects of energy; 1971–present
ENVIROLINE	Environmental Information Center, Inc.	international environmental information including management, planning, law, politics, as well as science and technology; 1971–present
EPB (Environmental Periodicals Bibliography)	Environmental Studies Institute	multidisciplinary index to environmental periodicals; 1973–present
EPICS (European Petrochemical Industry Computerized System)	Parpinelli-Tecnon, Italy	statistics and plant information
EUROLEX	European Law Center, Ltd., UK	full-text data base of the legal and business professions in Europe; anticipated in 1980
EXSTAT	Extel Statistical Services, Ltd., UK	international company information
FEDREG	Capitol Services, Inc.	citations from the U.S. *Federal Register;* multidisciplinary; 1977–present
FOREIGN TRADERS	U.S. Department of Commerce	directory of manufacturers, wholesalers, distributors, etc, in 130 countries that import or wish to import from the U.S.; a nonbibliographic file restricted to U.S. use; current 5 years
GPO MONTHLY CATALOG	U.S. Government Printing Office	reports, fact sheets, conference proceedings, etc, issued by all U.S. governmental agencies, including Congress; multidisciplinary; 1976–present
IBIS	CERVED, Italy	information on over 160,000 companies in about 130 countries
INTDB (International Data Base)	Chase Econometric Associates, Inc.	international nonbibliographic data base on commercial banks, government finance, international liquidity and transactions, and national income
IPC CHEMICAL DATA BASE	International Publishing Corporation, UK	international plant and product information for about 120 chemicals
ITIS	CERVED, Italy	economic outlook and commercial data on 90 countries

Table 4 (*continued*)

Data base name	Supplier	Coverage
KOMPASS-FRANCE	Kompass, and Société Pour d'Informatique, France	manufacturing companies, suppliers and products in France
LABORDOC	The International Labour Organization, Switzerland	international coverage of industry, economics, management, public finance, occupational safety and related fields from ILO publications; 1965–present
MAGAZINE INDEX	Information Access Corporation	multidisciplinary index to over 370 popular magazines; 1977–present
MANAGEMENT CONTENTS	Management Contents, Inc.	business and management related topics; 1974–present
METALS WEEK	McGraw-Hill	metal prices
NATIONAL NEWSPAPER INDEX	Information Access Corporation	front-to-back-page indexing of *The Christian Science Monitor, The New York Times,* and *The Wall Street Journal;* 1979–present
NEWSEARCH	Information Access Corporation	indexing of current newspapers, magazines, and periodicals; material is transferred to the NATIONAL NEWSPAPER INDEX and MAGAZINE INDEX at the end of each month; daily updates
NEW YORK TIMES INFORMATION BANK	New York Times Company	business, economics, politics, and law from major newspapers and magazines
PAIS INTERNATIONAL	Public Affairs Information Service, Inc.	international coverage of all fields of social science including political science, banking, public administration, law, policy, etc; 1976–present
P/E NEWS	American Petroleum Institute	extensive indexing of seven major periodicals in the petroleum and energy fields; 1975–present
POLLUTION ABSTRACTS	Data Courier, Inc.	citations to literature on pollution; 1970–present
PTS F&S INDEXES	Predicasts' Terminal System (PTS), Predicasts, Inc.	international company, product and industry information on acquisitions, mergers, new products or technology, socio-political factors, and summaries of forecasts; 1972–present
PTS FEDERAL INDEX	Predicasts Terminal Systems (PTS), Predicasts, Inc.	coverage of federal actions such as proposed regulations, bill introductions, contract awards, etc; 1976–present
PTS INTERNATIONAL ANNUAL TIME SERIES	Predicasts Terminal Systems (PTS), Predicasts, Inc.	time series including historical data from 1957 and projections for the future; all countries of the world are covered except the U.S.; economic, demographic, industrial, and product data; 1972–present
PTS INTERNATIONAL STATISTICAL ABSTRACTS	Predicasts Terminal System (PTS), Predicasts, Inc.	abstracts of published forecasts with historical data for all countries of the world except the U.S.
PTS PREDALERT	Predicasts Terminal System (PTS), Predicasts, Inc.	current month (records added on a weekly basis) corresponding to the cumulative files PTS PROMPT, PTS F&S INDEXES, and PTS FEDERAL INDEX

Table 4 (*continued*)

Data base name	Supplier	Coverage
PTS PROMPT	Predicasts Terminal System (PTS), Predicasts, Inc.	overview of markets and technology; marketing data for scientifically and technically oriented industries; current records in PTS PREDALERT; 1972–present
PTS US ANNUAL TIME SERIES	Predicasts Terminal System (PTS), Predicasts, Inc.	coverage as in PTS INTERNATIONAL ANNUAL TIME SERIES, but for the U.S. only; 1971–present
PTS US STATISTICAL ABSTRACTS	Predicasts Terminal System (PTS), Predicasts, Inc.	coverage as in PTS INTERNATIONAL STATISTICAL ABSTRACTS but for the U.S. only; 1971–present
QUEBEC-ACTUALITE	Microfor, Inc., Canada	French-language file of three Quebec newspapers, *Le Devoir, La Presse,* and *Le Soleil;* 1973–present
RAPRA ABSTRACTS	Rubber and Plastics Research Association of Great Britain	economic, commercial, technical, and research aspects of the rubber and plastics industries; 1972–present
SAFETY SCIENCE ABSTRACTS	Cambridge Scientific Abstracts	interdisciplinary coverage of safety literature; includes legislative regulations and their effect as well as technical aspects; 1975–present
SANI (Anagrafe Camerale ed Operativa delle Imprese Italiane)	CERVED, Italy	information on Italian companies
SGB (Société Générale de Banque de Belgique)	Société Pour d'Informatique, France	finance, administration, and management file developed by a major Belgian bank
SIBB (Archivio degli Indici dei Bollettini delle S.p.A. ed S.r.l.)	CERVED, Italy	financial information on registered Italian companies
TITLEX (Archivi dei Titoli della Legislazione Statale)	Centro Elettronico di Documentazione della Corte Suprema di Cassazione, Italy	titles of current Italian laws and decrees
TRIBUT (Archivio della Giurisprudenza della Commissioni Tributaria Centrale e delle Commissioni Tributarie di 1o e 2o Grado)	Centro Elettronico di Documentazione della Corte Suprema di Cassazione, Italy	Italian tax law
USPSD (United States Political Science Documents)	University of Pittsburgh	scholarly articles in the broad area of political science; 1975–1977
WORLD PETROCHEMICAL DATA BASE	Stanford Research Institute	international production information

(WP Data Base) covers approximately 80 petrochemicals and derivatives. A computer program allows complex searching stragegies and allows search results to be edited on line. Information in the WP Data Base is keyed to individual products and subdivided according to country. Within each country, product information is expressed in terms of manufacturing company, plants in operation during the period from 1974

Table 5. Balance for Ethylene in The Netherlands, in 1979, Thousand Metric Tons

Company	Location	Production	Consumption	Balance	Consumption breakdown				
					LDPE[a]	HDPE[b]	EO[c]	EDC[d]	EB[e]
Limbourg									
DSM	Beek	524	336	+188	267	69			
Area total		*524*	*336*	*+188*	*267*	*69*			
West									
Dow	Terneuzen	758	422	+336	160		80		182
Gulf	Rotterdam	242	53	+189					53
Shell	Pernis	113	47	+66			47		
Shell	Moerdijk	403	137	+266			113		24
Group total		*516*	*184*	*+332*			*160*		*24*
ICI	Rozenburg		125	−125	125				
AKZO	Botlek		177	−177				177	
Area total		*1516*	*961*	*+555*	*285*		*240*	*177*	*259*
other derivatives			16						
Country total		*2040*	*1313*	*+727*	*552*	*69*	*240*	*177*	*259*

[a] LDPE = low density polyethylene.
[b] HDPE = high density polyethylene.
[c] EO = ethylene oxide.
[d] EDC = ethylene dichloride.
[e] EB = ethylene dibromide.

to the current year, and forecast figures to 1985, captive use of product, and raw materials. The supply-and-demand section of the WP Data Base provides the year-end capacities, production years, imports and exports, and consumption figures. Ownership relations and subsidiary names are searchable too. An example from SRI's WP Data Base of a supply-and-demand report for one product covering North America is shown in Table 6.

Parpinelli-Tecnon's EPICS and SRI's WP data base allow rapid searching, analysis, consolidation, and editing of data that otherwise can only be laboriously assembled from the large volume of published material. Furthermore, current information is available promptly from the on line files; the latest published information may be several months old.

Other Business Data Bases. Several bibliographic files for business information have been characterized by J. L. Hall in *On-Line Bibliographic Data Bases—1979 Directory* published by ASLIB, London, Eng., or by A. Tomberg, *EUSIDIC Data-Base Guide 1980,* Learned Information, Abingdon, Eng. These provide references to other publications on on-line data bases, both bibliographic and nonbibliographic.

A few files are of particular interest to chemical business and marketing. References to published articles, and often full abstracts, on the chemical industry's activities and economic outlook may be found in the American Petroleum Institute's APILIT and in P/E NEWS, which provides technical-economic information for the petrochemical sector. These files have been described in ref. 32. The chemical industries are also covered by files such as Chemical Abstracts Service's CIN (Chemical Industry Notes) and by Predicasts' PTS PROMPT and PTS F&S INDEXES which supplement each other. In addition, Predicasts offers the EIS INDUSTRIAL PLANTS file which is an on-line directory of manufacturing industries providing for each plant the name and address of the operating company and its parent company, the principal product, the market share, total sales, the employment-size class, and size of shipments. This file is described in ref. 33 where Predicasts' PTS PROMPT, PTS F&S INDEXES, and the PTS statistical data bases are also described.

Information on companies that file reports with SEC in the United States can be searched on the file DISCLOSURE, which is accessible through Lockheed's DIALOG system. Information on companies registered in Europe can be searched

Table 6.   North American Supply and Demand for Low Density Polyethylene, Thousands of Metric Tons

Supply year[a]	Year-end capacity	Production	Imports	Exports	Apparent consumption	Consumption
1974	3242	2953	130	187	2896	2874
1975	3287	2415	60	151	2324	2312
1976	3465	2942	72	269	2745	2704
1977	4038	3270	119	286	3103	3077
1978	4264					3210
1979	4391					3390
1980	4947					3590
1981	5185					3801
1982	5480					4027
1983	5480					4270

[a] Data for 1980–1983 are estimated.

in the EXSTAT file. As of the end of 1979, the DAFSA LIAISONS FINANCIERES file, offered by the Société Pour d'Informatique (SPI), provides information on the financial relationships between companies. SPI also offers KOMPASS-FRANCE, which allows users to search for information on manufacturing companies, suppliers, and products in France, and a file developed by the Belgian bank, the Société Générale de Banque de Belgique (SGB), covering finance, administration, and management.

CISI in Paris offers the Analyse Financière sur Ordinateur (AFO) file which gives financial and stock-exchange data for French companies including five years of historical information.

The work of committees and subcommittees of the United States Congress is covered in the CIS INDEX data base, which includes industrial topics on both national and international governmental policies.

The Congressional Information Service developed the American Statistics Index (ASI), which is a key to all statistical publications of the United States government, eg, from the Bureau of Census, the Bureau of Labor Statistics, the National Center for Health Statistics, the National Center for Social Statistics, the Statistical Reporting Service, and the Department of Agriculture. The index also covers several statistical publications from other parts of the United States government.

CRECORD produced by Capitol Services, Inc., is an on-line index, with abstracts, of the *Congressional Record,* the official journal of the proceedings of the Congress. Capitol Services also developed FEDREG, the on-line file that provides references to the *Federal Register* on agriculture, business, economy, finance, environment, labor, resources, taxation, trade, and the like. The CIS INDEX, ASI, CRECORD, and FEDREG are all accessible on line through SDC's services.

Predicasts' PTS FEDERAL INDEX, accessible through Lockheed's DIALOG system, covers similar subjects from the *Washington Post,* the *Congressional Record,* the *Federal Register,* and *Commerce Business Daily.*

Information bases covering major dailies and periodicals are now available. A main feature of two of these, the NEW YORK TIMES INFORMATION BANK and NEWSEARCH, is their timeliness. The New York Times Company selects articles on business, economic, political, and legal topics from about 60 newspapers and magazines covering petroleum, petrochemicals, the chemical and allied industries. Some of the major publications which are covered include the *New York Times,* the *Wall Street Journal,* the *Washington Post, Business Week, Financial Times, Barron (Gadler and London), The Economist, Forbes, Fortune,* and *Journal of Commerce.* All front pages or main news articles are put on line within 24 h. This file is described in ref. 34.

NEWSEARCH, a daily reference file accessible via the DIALOG service, contains bibliographic records of news items and articles selected from publications such as the *Christian Science Monitor,* the *New York Times,* and the *Wall Street Journal* and over 370 popular American magazines. Information no longer current (more than 30 d old) is accessible from the MAGAZINE INDEX and NATIONAL NEWSPAPER INDEX files, which are updated monthly. These two files include such subject matter as company information, trademarks, trade names, legislation, business and finance, product evaluation, and current affairs whether national or international. The NATIONAL NEWSPAPER INDEX is accessible through the SDC and Lockheed services.

The number of business-information files is already considerable, and it grows as additional systems are computerized. Although the objective is to provide information promptly and effectively, the number of data bases and national languages will make it both easier in one sense and more difficult in another for information professionals to provide timely and comprehensive business data. Information professionals, while continuing to search several files for scattered bits and pieces of the puzzles they put together, will feel the need for a few comprehensive files concerned with the chemical sector that can bring together the information they need. Because such a utopian product is unlikely to appear in the next few years, business information will, without doubt, continue to be collected piecemeal from the wealth of sources worldwide.

BIBLIOGRAPHY

"Literature, Mechanized Searching" in *ECT* 1st ed., Vol. 8, pp. 449–467, by J. W. Perry, Bjorksten Research Laboratories, and R. S. Casey, W. A. Sheaffer Pen Co.; "Literature of Chemistry and Chemical Technology" in *ECT* 2nd ed., Vol. 12, pp. 500–529, by T. J. Devlin, Esso Production Research Co., and B. H. Weil, Esso Research and Engineering Co.; "Information Retrieval Services and Methods" in *ECT 2*, Suppl. Vol., pp. 510–535, by Eugene Garfield and Charles E. Granito, Institute for Scientific Information, and Anthony E. Petrarca, The Ohio State University.

1. D. F. Johnston, *Copyright Handbook*, R. R. Bowker Company, New York, 1978.
2. S. L. Burgoon, *The Viability and Impact of Electronic Storage and Delivery of Handbook-Type Information, PB 278072*, National Technical Information Service, Springfield, Va., 1978.
3. Y. S. Touloukian, *Bull. Am. Soc. Inf. Sci.* **5**(6), 44 (1979).
4. H. M. Woodburn, *Using the Chemical Literature: A Practical Guide*, Marcel Dekker, Inc., New York, 1974, pp. 111–123.
5. *Ibid.*, pp. 141–151.
6. A. Weissberger, ed., *Techniques of Chemistry*, Wiley-Interscience, New York, 1971–1979.
7. *Curr. Abstr. Chem.* **12**(3), (Jan. 17, 1979).
8. J. F. Terapane, *Chemtech* **8**(5), 272 (1978).
9. W. J. Bowman, *J. Chem. Inf. Comput. Sci.* **18**(2), 81 (1978).
10. H. M. Allcock and J. W. Lotz, *Chemtech* **8**(9), 532 (1978).
11. J. F. Terapane, *J. Chem. Inf. Comput. Sci.* **17**(3), 130 (1977).
12. S. M. Kaback, *Online (Weston, Conn.)* **2**(1), 16 (1978).
13. J. T. Maynard, *Chemtech* **8**(2), 91 (1978).
14. M. J. Marcus, *J. Chem. Inf. Comput. Sci.* **18**(2), 76 (1978).
15. K. M. Donovan and B. B. Wilhide, *J. Chem. Inf. Comput. Sci.* **17**(4), 139 (1977).
16. M. M. Duffey, *J. Chem. Inf. Comput. Sci.* **17**(3), 126 (1977).
17. R. G. Smith, L. P. Anderson, and S. K. Jackson, *J. Chem. Inf. Comput. Sci.* **17**(3), 148 (1977).
18. W. Marcy, ed., *Patent Policy: Government, Academic, and Industry Concepts, ACS Symposium Series 81*, American Chemical Society, Washington, D.C., 1978.
19. *Central Patents Index 1980*, Derwent Publications, Ltd., London, Eng., 1979.
20. R. J. Rowlett, Jr., *Chemtech* **9**(6), 348 (1979).
21. J. T. Maynard, *J. Chem. Inf. Comput. Sci.* **17**(3), 136 (1977).
22. A. Girard, S. M. Kaback, and K. Landsberg, *Database* **1**(2), 46 (1978).
23. S. R. Heller and G. W. A. Milne, *Environ. Sci. Technol.* **13**(7), 798 (1979).
24. M. Williams, *Computer-Readable Data Bases: A Directory and Data Sourcebook*, Knowledge Industry Publications, Inc., White Plains, N.Y., Oct. 1979; M. Williams, *Bull. Am. Soc. Inf. Sci.* **6**(2), 22 (1979).
25. R. N. Landau, J. Wanger, and M. C. Berger, eds., *Directory of On-Line Data Bases*, Cuadra Associates, Inc., 1980.
26. Union of International Associations (Brussels), ed., *Yearbook of International Organization 1978–1979*, 17th ed., International Publication Services, New York, 1978.
27. *Europa Yearbook, A World Survey*, 20th ed., 2 vols., Europa Publications, Ltd., London, Eng., 1979.

28. L. Huckfield, *Trade Ind.* **31**(9), 483 (June 2, 1978).
29. B. Niewenglowska, *ASLIB Proc.* **27**(11–12), 453 (1975).
30. M. Shollaert, *La Documentation Economique, Commerciale et Financière,* Colloque National de la Documentation, Association Belge de Documentation, Brussels, Belgium, May 9–10, 1974.
31. G. Bloch, *ASLIB Proc.* **27**(11–12), 459 (1975).
32. L. Rogalski, *J. Chem. Inf. Comput. Sci.* **18**(1), 9 (1978).
33. M. A. Burylo, *Online (Weston, Conn.)* **1**(3), 53 (1977); G. Sharp, *Online (Weston, Conn.)* **2**(1), 33 (1978).
34. D. R. Dolan, *Online (Weston, Conn.)* **2**(2), 26 (1978).

MARGARET H. GRAHAM
Exxon Research and Engineering Company

ALEXIS B. LAMY
Essochem Europe, Inc.

BARBARA LAWRENCE
Exxon Corporation

LORRAINE Y. STROUMTSOS
Exxon Research and Engineering Company

PATENTS, LITERATURE

Patent literature is a main source of technical and scientific knowledge. The information contained in patent documents is by definition new and often can be found nowhere else. In this era of rapid technological change, the chemist or engineer has to maintain a close watch on this massive body of information with regard to scientific development and innovation (see Information retrieval).

In general, a patent is a grant by a government to an inventor of the right to exclude others for a limited time from making, using, or selling the patented invention. However, the right to exclude others does not necessarily give the inventor the right to practice the invention. A patent is often granted on an improvement to a broader invention that could be the subject of an earlier patent. The broader patent could be infringed by practicing the improvement. Protection is granted only in the country issuing the patent. In exchange for this limited monopoly, the inventor is obligated to disclose his or her invention publicly in detail sufficient that someone skilled in the art may produce the same results. This provision for disclosure fulfills the patent's fundamental purpose, namely, the promotion of science and technology.

New technology is most often published first in the patent literature (1). A number of systematic studies have been directed at determining the gap existing between the patent and nonpatent literature as sources of technical information. In a study conducted by the Office of Technological Assessment, U.S. Patent and Trademark Office, the new technology identified in a random sample of 435 patents issued in 1967 and 1972 was examined for disclosure of the same technology in nonpatent publications. The results indicated that $71 \pm 10\%$ of the patent information is not reported in the general literature (2).

Because a patent document represents a legal contract that requires certain form and style, many chemists and chemical engineers regard the patent literature with a somewhat negative attitude. The objective of a patent disclosure is different from that of a journal article; once the reasons for this difference are understood, however, the significant information contained in the patent can be identified and utilized.

Patent Documents

Variations in patent laws from one country to another have given rise to multiple forms and kinds of patent documents. The first printed document is often a published unexamined application. In the United States, however, an application is held confidential until a decision to grant the patent has been made. Only then is publication initiated. Changes to already-issued patents are handled differently from country to country. In the United States, some types of errors are taken care of through a certificate of correction, but others may result in a reissue patent numbered in a different, five-digit series. In the United Kingdom, an amended specification is so labeled, but not renumbered. Many countries allow for addition or improvement patents, the terms of which generally expire with the main, principal, or independent patents to which they relate. Some countries allow for dependent patents, the terms of which are separate; however, such a patent cannot be worked without infringing the main patent while it is still in force. Although most countries number addition patents in the regular patent series, French old-law addition patents are numbered in a different series. When a main patent is invalidated or revoked, an addition or improvement patent usually becomes a main patent. Where there is no special provision for such patents (as in the United States), the new feature, if sufficiently inventive, is simply patented as a separate invention. Some countries also provide for patents of confirmation, revalidation, importation, or registration to extend the protection of a foreign patent to the country concerned. The petty patent or utility model (German *Gebrauchsmuster*) (3) is designed to give protection to discoveries whose inventiveness is insufficient to justify a regular patent. It is important in the FRG and Japan, and, for a different reason, in Spain where it is the only way to protect an article of manufacture. Author's and inventor's certificates (used mostly in USSR-bloc countries) have the documentary appearance of a patent; however, their function is that of an official, publicized receipt for patent rights granted to the state (4). The author's certificate, which is mandatory for all inventions arising from employment by the state, bears a formal resemblance to employer–employee patent-rights contracts in other countries. No fees are charged the inventor for prosecution of the application. If successful, remuneration is provided in accordance with any savings made by the state through use of the invention. The GDR industrial (or economic) patent is similar, except that the patentee retains nonexclusive rights and is charged half of the fee for a regular or exclusive patent. Design patents protect only the external appearance of an article such as an instrument panel, toy, or mechanical device (5). Although commercially important in many countries, they, like patents on new strains of plants produced in certain specified ways (plant patent) (6), are of a different nature from patents for invention and are not discussed here. To further confuse the casual patent inquirer, in addition to the multiplicity of patent types available, several levels of publication, with respect to the same application, may occur. In the Federal Republic of Germany, for example, a patent document may be published at each of three procedural stages resulting in an Offenlegungsschrift, an Auslegeschrift, and a Patentschrift. More information on the patent documents of specific countries is available from the respective national patent office. Most offices are very cooperative in supplying information about their services. A list of patent-office addresses may be found in a number of publications one of which is the *International Table to Patents, Designs, and Trade Marks*, a wall chart (7).

Standardization. Fortunately, procedures encouraged under the Patent Cooperation Treaty and administered through the International Bureau of the World Intellectual Property Organization (WIPO) have brought a level of order and standardization to patent documents. The Paris Union Committee for International Cooperation in Information Retrieval Among Patent Offices (ICIREPAT) has developed four standards of significance for the handling and use of patent information. First, they have provided a two-letter standard code system for identifying states and organizations which facilitates the consistent entry of country information in computer storage and retrieval systems. A list of the codes is given in Table 1.

The second ICIREPAT standard of importance to the information scientist is the standard code for identification of different kinds of patent documents. This code provides groups of single letters to distinguish the various types of patent documents and the various levels of publication. In addition, any patent office desiring to identify subdivisions unique to its documents may provide a numerical suffix. These codes are used primarily in machine processing of patent information. It also has been recommended that they appear on the first page of patent documents and as identification for references in patent gazettes. The code is shown in Table 2. Group 1, or the main series, refers to regular patents of invention. For example, the three levels of documents resulting from a Netherlands (NL) application would be identified as follows: Terinzagelegging (patent application laid open)—A, Openbaarmaking (published examined application)—B, Octrooi (published patent)—C. Group 2, or the secondary series, refers to patent documents of a secondary nature such as the U.S. reissue and French patent of addition. Group 3 is reserved for special patent documents not classifiable under Group 1 or 2. U.S. defensive publications are an example of this category. Group 4 refers to a main-series patent that covers a special subject area. French medicament patents and U.S. plant patents are examples. The fifth group is for utility models and the sixth for nonpatent and restricted documents.

The third standard, to which most Paris Union members adhere, provides that each patent document have a single front page containing a title, bibliographic and filing information, an abstract, and, if appropriate, a drawing. This allows the user to scan the front page and quickly determine the relevance of the document to the search at hand. A fourth standard provides codes for identifying the various bibliographic-data elements present on the front page. Regardless of the language of the document, this code enables the user to identify quickly desired data. It is particularly useful in ascertaining filing and priority dates. A list of the INID codes (ICIREPAT Numbers for Identification of Data) is given in Table 3.

Dates. Several dates are important in the life of a patent: the filing or application date; the date when the document is first opened to public inspection (often abbreviated OPI); the date when it is published (printed copies available); the date when the application is accepted (U.S. allowed) by the patent office as among all statutory

Table 1. Code for Identifying States and Organization

Code	Country	Code	Country
AR	Argentina	LK	Sri Lanka
AT	Austria	LU	Luxembourg
AU	Australia	LY	Socialist People's Libyan Arab Jamahiriya
BE	Belgium	MA	Morocco
BG	Bulgaria	MC	Monaco
BJ	Benin (Dahomey)	MG	Madagascar
BR	Brazil	MR	Mauritania
BS	Bahamas	MT	Malta
CA	Canada	MU	Mauritius
CF	Central African Empire	MW	Malawi
CG	Congo	MX	Mexico
CH	Switzerland	NE	Niger
CI	Ivory Coast	NG	Nigeria
CM	Cameroon	NL	Netherlands
CS	Czechoslovakia	NO	Norway
CU	Cuba	NZ	New Zealand
CY	Cyprus	PH	Philippines
DD	German Democratic Republic	PL	Poland
DE	Germany, Federal Republic of	PT	Portugal
DK	Denmark	RH	Southern Rhodesia
DO	Dominican Republic	RO	Romania
DZ	Algeria	SE	Sweden
EG	Egypt	SM	San Marino
ES	Spain	SN	Senegal
FI	Finland	SR	Surinam
FR	France	SU	Soviet Union
GA	Gabon	SY	Syrian Arab Republic
GB	United Kingdom	TD	Chad
GH	Ghana	TG	Togo
GR	Greece	TN	Tunisia
HT	Haiti	TR	Turkey
HU	Hungary	TT	Trinidad and Tobago
HV	Upper Volta	TZ	United Republic of Tanzania
ID	Indonesia	UG	Uganda
IE	Ireland	US	United States
IL	Israel	UY	Uruguay
IQ	Iraq	VA	Vatican City State (Holy See)
IR	Iran	VN	Viet Nam, Socialist Republic of
IS	Iceland	YU	Yugoslavia
IT	Italy	ZA	South Africa
JO	Jordan	ZM	Zambia
JP	Japan	ZR	Zaire
KE	Kenya		
LB	Lebanon	OA	African Intellectual Property Organization
LI	Liechtenstein	EP	European Patent Office
		WO	International Bureau of WIPO

Table 2.   ICIREPAT Code for Identification of Patent Documents

Group	Purpose	Code	Publication level	Examples
1	primary or main series	A	first	U.S. patent
		B	second	FRG auslegeschrift
		C	third	FRG patentschrift
2	secondary series	E	first	U.S. reissue
		F	second	
		G	third	
3	further series for special requirements	H		U.S. defensive publication
		I		
4	principal special types	M		medicinal
		P		plant patents
5	utility models[a]	U	first	FRG gebrauchsmuster
		Y	second	
		Z	third	
6	others	N		nonpatent literature
		X		restricted for internal use

[a] With a numbering series other than documents of group 1.

requirements; the date the patent is granted (U.S., issued; UK, sealed); and, of course, the dates the patent term comes into force and expires. In addition, two other important dates require more detailed comments. Many countries use an opposition period, usually two to four months, during which the text of the examined application is laid open to the public or, in a few important countries, issued in printed form. Interested parties can then file with the patent office any reasons for thinking the patent should not be granted. In some countries (FRG), this means anyone; in others (UK), only someone with a *bona fide* commercial interest can oppose granting the patent. Hence, the length or expiration date of the opposition period is important.

The International Convention for the Protection of Industrial Property stipulates that any patent application filed in a convention country within one year from the date of the earliest filing of the same invention in another convention country has all the legal benefits of the earlier filing with respect to priority. Hence, this earliest foreign filing date, which is called the priority date or convention date, is the most important date in an application in a convention country. The situation is sometimes complicated by the use of partial or multiple priorities that arise when two or more related applications filed in a convention country are the basis for a single application in another country. This is possible because of differing ideas as to what constitutes a single invention (unity of invention).

Document Numbers.   All patent offices assign an application or filing number (U.S. serial number) to applications as received. After the patent is granted, it bears a different patent number. When patents are published before grant (eg, to invite opposition), they usually bear either their application number or a new number, which will be the patent number if the potential patent is actually granted. A third alternative is to use a separate number especially for specifications accepted and published for opposition, as in the Netherlands and Japan. This is variously called publication number, accepted specification number, or published application number. Because Japanese examined and unexamined applications do not always carry a unique number, it is necessary to note the publication level to distinguish between them.

Table 3. INID Codes and Subcodes[a]

[10] Document identification
  [11] number of the document
  [19] ICIREPAT country code or other identification of the country publishing the document
[20] Domestic filing data
  [21] number(s) assigned to the application(s)[b]
  [22] date(s) of filing application(s)
  [23] other date(s) of filing, including exhibition filing date and date of filing complete specification following provisional specification
[30] Convention priority data
  [31] number(s) assigned to priority application(s)
  [32] date(s) of filing of priority application(s)
  [33] country (countries) in which priority application(s) was (were) filed
[40] Date(s)
  [41] date of making available to the public by viewing, or copying on request, an unexamined document, on which no grant has taken place on or before the said date
  [42] date of making available to the public by viewing, or copying on request, an examined document, on which no grant has taken place on or before the said date
  [43] date of publication by printing or similar process of an unexamined document, on which no grant has taken place on or before the said date
  [44] date of publication by printing or similar process of an examined document, on which no grant has taken place on or before the said date

[45] date of publication by printing or similar process of a document, on which grant has taken place on or before the said date
[46] date of publication by printing or similar process of the claim(s) only of a document
[47] date of making available to the public by viewing, or copying on request, a document on which grant has taken place on or before the said date
[50] Technical information
  [51] international patent classification
  [52] domestic or national classification
  [53] universal decimal classification
  [54] title of the invention
  [55] key words
  [56] list of prior-art documents, if separate from descriptive text
  [57] abstract or claim
  [58] field of search
[60] Reference(s) to other legally related domestic document(s)
  [61] related by addition(s)
  [62] related by division(s)
  [63] related by continuation(s)
  [64] related by reissue(s)
[70] Identification of parties concerned with the document
  [71] name(s) of applicant(s)
  [72] name(s) of inventor(s) if known to be such
  [73] name(s) of grantee(s)
  [74] name(s) of attorney(s) or agent(s)
  [75] name(s) of inventor(s) who is (are) also applicant(s)
  [76] name(s) of inventor(s) who is (are) also applicant(s) and grantee(s)

[a] ICIREPAT numbers for data identification.
[b] For example, Numero d'enregistrement national or Aktenzeichen.

PRIMARY INFORMATION

Primary-source access to patent information is through the more than 130 national and regional patent offices throughout the world. All Paris Union member states must publish a periodic journal (frequently called a gazette) listing the patent owner and providing an abstract for each patent granted. Most countries provide printed copies of applications or granted patents at prescribed procedural stages during the prosecution of an application. The principal offices also publish annual indexes, usually including a listing of patents by applicant or assignee and by inventor, as well as other summaries and notices.

The timing and format of information emanating from the world's patent offices varies greatly. Since U.S. patent law requires that applications be held in confidence until a patent is granted, publication can take anywhere from six months (very unusual) to two (normal), three, or even five years. Multiple printed copies then become available. The procedure is different in other countries. In Belgium, for example, patents are granted and laid open for public inspection (OPI) three and one-half to six months after the filing date. Printed copies, however, are not available at that time. Patent applications may be inspected as soon as filed in Panama and in Peru. In other Latin American countries, applications may be inspected a few weeks after filing (8). Beyond this, the waiting period ranges from several months to over ten years for access to a given specification. A few countries (United Kingdom, Australia, South Africa, Italy, Belgium) publish brief titles of new applications in their official journals a few weeks after filing. However, the brief titles and other filing data thus made available contain so little information that they are useful only under special circumstances. Portugal is unique in publishing the complete claims of applications in its official journal at the end of the month following the month in which the application was filed. The specification, however, is kept secret until three years after grant, which generally occurs seven to fourteen months after filing. Patent-office journals and gazettes also are the source of other kinds of patent-related information, much of which is not available elsewhere. In countries where opposition is allowed, notification of this information frequently is provided. Where maintenance fees are required, lapses are noted. Termination of patent rights for other reasons also are listed; for example, disclaimers, withdrawals, and invalidations. Gazettes are published on various schedules, depending upon the rate of issue in the particular country. Most main patent-issuing countries publish their gazettes weekly, semimonthly, or monthly. The less-prolific offices publish less frequently; some, at irregular intervals. Specific information about particular patent gazettes may be obtained by writing to the appropriate national patent offices.

To satisfy most information needs, a printed copy of the full document is desirable. The timing of publication varies greatly from country to country (9). The publication date depends mainly on the examination procedure used by the granting state. Most industrialized countries subscribe to the concept of examining alleged inventions for novelty. In some countries, the United States for example, all applications for invention are examined for novelty; in others, such as Japan and the Netherlands, examination is provided upon request of the applicant or a third party. Portugal examines for novelty only if the application has been opposed (a procedure whereby a third party may file an opposition to the granting of a patent within a certain period of time following public notification). The requirement for novelty varies from absolute (prior publication anywhere defeats the application) to domestic only. The examination itself also varies from a search through domestic patents or through domestic literature in general, to a thorough search using all available world literature. Mexico normally searches only domestic publications and makes an absolute novelty search on request. In many countries the degree of inventiveness or unobviousness of the invention is assessed, and a suggestion of at least some utility is required. The Federal Republic of Germany requires in addition a proof of technical progress, which means that the invention cannot be simply another method or device for some purpose, but must be a better one (cheaper, higher quality, etc). USSR examiners, since they represent not only the patent office but also the assignee (the state), in the case of author's certificates are most particular about utility. All patent offices inspect new applications to see that at least the formal requirements have been met (fees paid, sworn statements, matters of format, subject matter in a patentable category, and often utility of invention). This is all the examination that some countries such as Italy and Spain require. They are the nonexamining or registration countries.

Although no two countries seem to operate in exactly the same manner, most can be grouped into one of the four generalized examining procedures shown in Figure 1. A list of the countries which, in general, follow each of the four procedures is given in Table 4. Most procedure-4 countries allow access to the specifications shortly after filing. All procedure-3 countries publish or allow access to specifications 18 months after filing. These are considered the fast-issuing countries. The countries following procedures 1 and 2 require complete examination for novelty before publication and accordingly are the slow-publishing countries. Although turnaround time for U.S. patents has improved during the last six to eight years, the average lag time between application and publication is still around two and one-half years. Figure 2 shows the average time between the filing and issue or publication dates for chemical patents issued the first weeks of January, April, July, and October in the years 1959, 1964, 1969, 1974, and 1979. The new fast-publication procedure (procedure 3) is being followed by more and more countries. It involves publication of the unexamined application about 18 months after filing and novelty examination upon request of the applicant or a third party, or in some cases, at the filing of an opposition. The traditional examination–acceptance–opposition–grant procedure (procedure 2) is still followed by

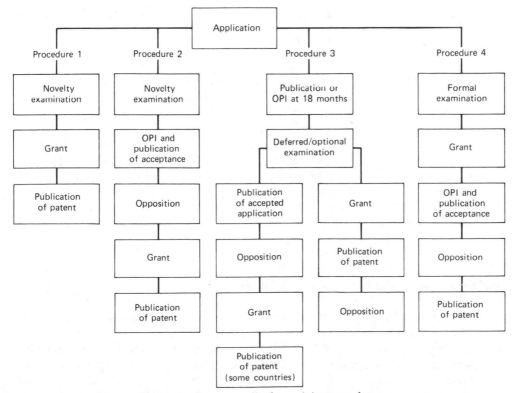

Figure 1. Four generalized examining procedures.

Table 4. Countries that Follow the Four Generalized Examining Procedures

Slow publishing		Fast publishing	
Procedure 1	Procedure 2	Procedure 3	Procedure 4
examination with publication after grant	examination with publication or OPI[a] of examined document and opposition period	publication of unexamined application; optional or deferred examination	examination for form only; OPI[a] or publication soon after filing
Argentina	Australia	Brazil	Belgium
Canada	Austria	Denmark	Italy
United States	Columbia	EPO[a]	Luxembourg
USSR	Czechoslovakia	Finland	South Africa
	GDR	France	Spain
	Hungary	FRG[b]	Switzerland[c]
	India	Japan[b]	Venezuela
	Ireland	Mexico	
	Israel	Netherlands[b]	
	New Zealand	Norway	
	Pakistan	Poland	
	Romania	Portugal	
	Switzerland[c]	Sweden	
	Yugoslavia	United Kingdom[b]	
		WPO[a]	

[a] OPI = opened to public inspection; EPO = European Patent Office, WPO = World Patent Organization.

[b] Recent law changes provide for continued examination of applications filed prior to the date of the new law under procedure 2. The issuance of old-law patents will eventually cease for these countries.

[c] Applications in area of chronology and textile treatment are examined, others are not.

a number of countries that have not yet changed to the fast-publishing system. Publication comes when the novelty examination has been completed, and the application has been accepted, rather than automatically 18 months after filing. Otherwise, the procedures are similar. The simplest examining procedure is the registration method whereby the issuing office limits its examination to the formalities mentioned above. In countries using this method, the merit or patentability of the alleged invention and the validity of the resulting patent is left for the courts to decide at a later time. The question: What good is a patent which is simply registered and thus without any presumption of validity? is discussed in ref. 10. Belgium is an example of a country using the nonexamining procedure.

The United States Patent and Trademark Office

The organization responsible for administering the provisions of U.S. patent law is the United States Patent and Trademark Office (PTO), an agency of the U.S. Department of Commerce. As its name implies, its mission is to provide patent protection for inventions and a registration system for trademarks. Copyrights are now registered by the copyright office at the Library of Congress and are no longer a responsibility of the PTO.

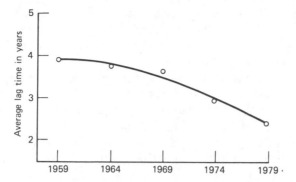

Figure 2.   Average lag time between application and grant for U.S. chemical patents.

The PTO grants four kinds of patents: patents (of invention), reissue patents, plant patents, and design patents. It also registers and publishes a patentlike document called a defensive publication. Some 60,000 patents issue each year; numbered consecutively, they are currently in the four-million (4×10^6) range. The reissue patent is issued when errors of substance are detected in a patent as granted. It refers to the original patent and delineates the added and deleted material. About 300–400 of these issue annually; they, too, are numbered consecutively and are currently in the 30,000 range. Plant patents are granted to anyone who has invented or discovered and asexually reproduced a distinct and new variety of plant. A design patent is granted to anyone who has created a new, original, and ornamental design for an article of manufacture. A design patent differs from a regular patent in that it protects only the appearance of an article, not its structure or utility. Occasionally, when patentability is questionable or the economics are not attractive, it becomes undesirable to file a patent application or to continue the prosecution of one already filed. In order to preclude the possibility of someone else filing for a patent on the same subject matter later when the economics may have improved or the utility of the invention increased, it is desirable to disclose the invention and in this way provide anticipatory prior art. The PTO provides a mechanism for this by registering and publishing a patentlike document called a defensive publication. Although it resembles a patent in form and is cited in the *Official Gazette*, it has no claims and no protective force in the sense of a regular patent. It serves the same purpose as any other published literature reference.

U.S. patent law requires that an invention, to be patentable, must be new, unobvious, and useful. Each application is critically examined on the basis of these criteria before it is either rejected or accepted. This process can take as long as three or four years. The present average for chemical patents is about 29 months. During this period, the application is held in strict confidence. Only after a patent has been granted is its content made public. Patents are issued on Tuesday of each week at which time printed copies of the patent specifications become available. The term of a U.S. patent is 17 years from the issue date. Printed patent documents may be obtained from the PTO on an individual basis by number or through an annual subscription by U.S. Patent Office Classification. They also may be purchased from Rapid Patent International (2221 Jefferson Davis Highway, Arlington, Va., 22202) individually by number or on a subscription basis by class or assignee.

The Official Gazette. *The Official Gazette of the United States Patent and Trademark Office* (OG), is published on each Tuesday and carries, for each newly issued patent, bibliographic data and one or more of the claims. In addition, it contains the following items of interest from the literature point of view: reissue applications filed (unlike patent applications which are held confidential, reissue applications are open to public inspection); certificates of correction (list of patents for which minor corrections have been filed); disclaimers and dedications (list of patents including inventor's name, assignee, title, and date of issue, for which a disclaimer or dedication has been entered); National Technical Information Service (NTIS) notice (list of U.S. government-owned patents and applications available for licensing); list of depository libraries (includes addresses and telephone numbers); index of patentees (arranged by inventors and cross-referenced by assignee); index by classification; and geographical index of residence of inventors (arranged by U.S. states, territories, and armed forces). The *Official Gazette* is available from the PTO on an individual-copy or annual-subscription basis.

The PTO provides a number of additional publications useful to those seeking patent information. Some of these are sold by the PTO; others are available from other agencies. A summary of these is given below.

Manual of Classification. Printed copies of schedules that give the class and subclass numbers and titles (but not patent numbers) together with an alphabetical index of subject matter are sold as the *Manual of Classification of Patents* and can be obtained from the Superintendent of Documents, Government Printing Office, Washington, D.C. This manual is available in the patent and trademark office search room in Arlington, Virginia, and in many libraries across the country.

Technical Bulletins, Class Definitions. These bulletins, also available in the Patent Search Room and in many libraries, supplement the *Manual of Classification* by providing definitions and search notes describing and illustrating the kinds of information or patents that can be found in the individual classes and subclasses. They are identified by the class they describe. Copies are sold by the PTO at prices based on their size. They can be ordered from the Office of Planning Support and Control, U.S. Patent and Trademark Office, Washington, D.C., 20231.

Microfilm Listings. PTO has various classification listings and changes, available on microfilm through the National Technical Information Service, Springfield, Va.

Annual Indexes of Inventors and Assignees. The PTO indexes newly issued patents each fiscal year by the name of inventor and by the name of the assignee of record at the time of issue. The compilation is printed as the *Index of Patents*. Commencing with the volume for 1955, it also contains a list of patent numbers issued during the year, arranged in sequence by the class and subclass number in which the patents were classified at the time of issue. Copies are available in public libraries of the larger cities. It is sold by the Superintendent of Documents.

Subclass Lists. These are lists of patent numbers arranged by subclass codes. Original referenced and cross-referenced patents are listed separately. Copies of either listing for any specified subclass may be ordered from the PTO. They may be used, for example, in conducting a class search in numerical files such as those maintained by most of the depository libraries.

The following pamphlets and booklets also are available from the Superintendent of Documents, U.S. Government Printing Office, Washington, D.C., 20402.

General Information Concerning Patents. This pamphlet, designed for the layman, contains a large amount of general information expressed in nontechnical language concerning the granting of patents.

Patents and Inventions, an Information Aid for Inventors. This pamphlet offers help to inventors in deciding whether to apply for patents, how to obtain patent protection and promote their inventions.

Patent Attorneys and Agents Registered to Practice Before the U.S. Patent Office. This list of registered attorneys and agents is arranged alphabetically and by states and countries.

The PTO maintains a public-search center at Crystal Plaza, 2021 Jefferson Davis Highway, Arlington, Va., 22202. Although the PTO's primary purpose in providing patent-searching accommodations is to satisfy the needs of its examining staff, significant search facilities also are provided for use by the public. The Patent Search Center contains over 4×10^6 U.S. patent documents granted since 1836 arranged by U.S. patent classification. Where a patent discloses subject matter provided for in two or more subclasses, a copy is placed in each subclass. To distinguish these copies from each other, the copy placed in a subclass selected as the primary basis of classification is called an original reference and all others are called cross-references. In addition to the classified patent collection, over 120,000 volumes of scientific and technical books are available in the search room, along with 90,000 bound volumes of scientific periodicals. Over 8×10^6 foreign patents, some searchable by national-classification schemes, and the patent journals of many foreign offices are also held there. A record room at the search center houses a complete set of U.S. patent documents bound in numerical order and a complete set of the *Official Gazettes*. Also available for public inspection are the files on granted patents. These are known as file histories and contain all the documentation created during the prosecution of the application. Copies of the file histories may be ordered at the search center or through Rapid Patent International.

In order to accommodate those who cannot visit the public-search facilities at Arlington, Virginia, the PTO provides copies of newly issued patent documents to a number of libraries throughout the country. These are known as patent-depository libraries. The scope of these collections varies from library to library, ranging from patents of only recent months or years in some libraries to all or most of the patents issued since 1870, or earlier, in other libraries. These patent collections are open to public use. Each of the patent-depository libraries offers, in addition, the publications of the patent-classification system (eg, *The Manual of Classification*, *Index to the U.S. Patent Classification*, *Classification Definitions*, etc) and provides technical staff assistance in their use. A list of the depository libraries is published in the *Official Gazette*. With one exception, the collections are organized in patent-number sequence and are not searchable by subject matter. The only library providing access by classification is in Sunnyvale, California.

SECONDARY SOURCES

Because of the worldwide nature of patent literature and the accompanying language and procurement problems, secondary abstracting and indexing services and custom searching play an important role in the dissemination of patent information. Most secondary services use computer-assisted systems in order to handle the increasing volumes of data now emanating from the world's patent offices (see Computers). Application of modern data-processing techniques to patent information has made it possible to keep pace quickly and economically with this ever-growing record of science and technology. A few years ago it was unthinkable that within a few weeks of publication bibliographic information on almost a million (10^6) newly published patent documents annually could be made available for instant recall almost anywhere in the world. This is now being done routinely in various forms through worldwide on-line communication networks. Chemical Abstracts Services, INPADOC, Derwent, and IFI/Plenum are the principal producers of on-line patent data bases. These on-line files together with the printed services offered by these and other companies and by national patent offices and societies as well as the custom searching offered by countless commercial and government-sponsored organizations provide a tremendous store of readily available technological information.

On-Line Services

Commercial on-line interreactive information searching was developed during the mid-1960s. Data stored at a remote computer facility are searched through the use of a keyboard terminal connected to the computer by a telephone link. The inquirer issues instructions to the computer by way of the keyboard and a set of natural-language commands. Immediate (or almost immediate) response to a command is received through the terminal, allowing the user to react to the results of one command before proceeding to the next. In this way the inquirer may alter search strategy and investigate new approaches suggested by the previous answers. In a sense, this provides a technique for browsing through the file. On-line services require the integration of three independently provided facilities: the storage and retrieval system and hardware, the telephone communications network, and the data base (a machine-readable collection of data referencing each of the represented documents).

In the United States, two major networks, Telenet and Tymnet, offer the communications link. Lockheed Missiles & Space Company (LMSC), System Development Corporation (SDC), and Bibliographic Retrieval Services, Inc. (BRS) provide most of the retrieval systems capability. Together, SDC and LMSC provide access to over 125 different data bases supplied to them by independent data-base producers. The data bases cover science and technology, social sciences and humanities, and business and economics. Many of them include patent information. The most important ones are listed in Table 5.

The Lockheed DIALOG system was initially developed for NASA during the 1960s (11). The version used by NASA is known as the NASA/RECON system. The present DIALOG system is a completely redesigned version with greatly expanded capabilities. The System Development Corporation service uses the ORBIT system which evolved during the same period from a developmental system called Protosynthex. Although these two services are similar in concept, much to the consternation of the user, they are somewhat different in application. Many of the logic capabilities, formats, and command languages are unique to each system and require the user to master both (12). On-line data bases are most frequently charged for by the hour. Where the data base evolved from a printed service and the continuing costs for updating are high, an annual subscription fee for the printed service is also charged for. Today, most corporate and academic and many public libraries have terminal

Table 5. On-Line Data Bases Containing Patent Information

File identification	Starting date	Subject coverage	Country coverage[a] data base supplier	Retrieval system[b] LMSC	SDC	BRS
Primary patent files						
APIPAT	1964	petroleum refining, petrochemicals	9 countries, EPO and WPO		X	
Chemical Abstracts	1967	chemical	6–17 countries, partial coverage of 11 more	X	X	X
Derwent CPI	1963	pharmaceuticals	12–15 countries, partial coverage in 11 more		X	
	1965	agriculturals				
	1966	polymers				
	1970	remaining chemical				
Derwent WPI	1974	nonchemical	as in Derwent CPI but lacking Japan		X	
Derwent EPI	1980	electrical/electronics	includes Japan		X	
IFI CLAIMS						
bibliographic	1950	chemical	United States, plus equivalents in 5 others	X		
	1963	nonchemical	United States	X		
Uniterm Index	1950	chemical	United States, plus equivalents in five others	X		
Class		U.S.P.O. Classification		X		
Citation	1947	all	United States	X		
INPADOC IPG	c	all	47 countries, EPO and WPO	X		
Pergamon PATSEARCH	1971	all	United States			X
Pergamon PATCLASS	1836	all (class index only)	United States			X
Other on-line files						
APTIC	1966[d]	air pollution	U.S. Environmental Protection Agency	X		
CAB Abstracts	1973	agriculture, biology	The Commonwealth Agricultural Bureaux, UK	X		
COLD		cold regions			X	
FSTA	1969	foods	International Food Information Service, UK	X	X	
Foods Adlibra	1974	foods	K & M Publications, Inc.	X		
Forest		wood products			X	
Geoarchive	1969	geoscience	Geosystems, England	X		
INSPEC	1969	physics, electrical, electronics, computers	The Institution of Electrical Engineers, UK	X	X	X
IRL Life Sciences	1978	life sciences	Information Retrieval, Ltd., England	X	X	
NTIS	1964	government sponsored research	National Technical Information Center	X	X	X
Paperchem	1968	pulp, paper	Institute of Paper Chemistry		X	
Pollution Abstracts	1970	pollution	Data Courier, Inc.	X	X	X
RAPRA Abstracts	1972	polymers, rubbers, plastics	Rubbers and Plastics Research Association of Great Britain	X		

Table 5 (*continued*)

File identification	Starting date	Subject coverage	Country coverage[a] data base supplier	Retrieval system[b] LMSC	SDC	BRS
SCISEARCH	1974	science	Institute for Scientific Information	X		X
Surface Coating Abstracts	1976	coatings	Paint Research Association of Great Britain	X		
TITUS	1970	textiles	Institute Textile de France		X	
TULSA	1965	petroleum exploration, production, transport	University of Tulsa		X	
WELDASEARCH	1967	welding	The Welding Institute, UK	X		
World Aluminum Abstracts	1968	aluminum	American Society for Metals	X		
World Textiles	1970	textiles	Shirley Institute, UK	X		

[a] Country coverage for Primary Files and Database supplier for Other On-line Files. EPO = European Patent Office; WPO = World Patent Organization.

[b] LMSC = Lockheed Missiles & Space Company, Inc. SDC = System Development Corporation. BRS = Bibliographic Retrieval Services, Inc.

[c] Most recent six weeks.

[d] Discontinued in September, 1978.

facilities making access to on-line files readily available. Although on-line files vary considerably in coverage, completeness, and timeliness (13–14), they are nevertheless an excellent source for alerting, preliminary anticipation, and assignee and inventor searching. Up-to-date listings of the data bases offered and other details on using these files may be obtained by writing to the retrieval system companies.

A source of information on patents deriving from U.S. Government-sponsored research is the RECON data base of the Department of Energy.

Chemical Abstracts Service

Chemical Abstracts (CA), published weekly through the Chemical Abstracts Service (CAS) of the American Chemical Society, provides an excellent abstracting and indexing service for new chemical information. Patent coverage, however, is not as broad as that found in the Derwent and IFI/Plenum files. Patents that are borderline with respect to providing new chemistry and particularly patents on chemical manufactures and applications are frequently not abstracted. For example, in 1978, CAS abstracted or cited as equivalents 14,617 U.S. patents (15), whereas according to U.S. patent classification, 21,792 chemical patent documents were issued in that year (16). Coverage of foreign patents has also not always been complete. In 1955, CAS covered all chemical patents from France, Federal Republic of Germany, Great Britain, and the United States and selected patents issued to nationals from 14 additional countries. Today, coverage has been expanded to all chemical patents from 15 countries and two regional offices plus selected patents issued to nationals in 11 other countries (see Table 6). Patent abstracts are placed into one of the 80 sections of CA on the basis of their overall subject content. They appear as a group at the end of each section. Indexing

Table 6.   Countries covered in *Chemical Abstracts*

All chemical patents	Selected patents issued to nationals
Austria	Australia
Belgium	Czechoslovakia
Brazil	Denmark
Canada	Finland
European Patent Organization (EPO)	German Democratic Republic
France	(GDR)
Federal Republic of Germany (FRG)	Hungary
Israel	India
Japan	Norway
Netherlands	Poland
Romania	Spain
South Africa	Sweden
Switzerland	
United Kingdom	
United States	
USSR	
World Patent Organization (WPO)	

is selective. Emphasis is placed on reporting new exemplified substances. In 1979, indexing policy was expanded to include new substances mentioned in the claims as well as those exemplified. Disclosures that are judged prophetic are not indexed. In spite of these limitations, since most chemists and chemical engineers are familiar with *Chemical Abstracts* and since it is so readily available at most corporate and academic libraries, it is an excellent place to start a search and is the mainstay of any state-of-the-art search (17).

Formerly, corresponding or equivalent patents were handled by making a regular entry referring the reader to a prior entry containing the abstract. Since 1963 this system has been replaced by a separate computer-produced patent concordance that lists all patents received that correspond to previously abstracted specifications, indicating the number of the patent abstracted and its location in CA. Under the number of the patent actually abstracted, numbers of all corresponding patents encountered to date are also given. In 1978, 57,000 patents were abstracted and 70,000 equivalents cited. In 1981, the numerical patent index and the patent concordance are being replaced by an expanded patent index. The new index is published in the weekly issues of CA and in the several cumulative indexes. It contains references to new documents abstracted during the period as well as equivalent patents abstracted previously. Data about related documents, eg, U.S. divisions, related nonpriority documents, and certain information with regard to level of publication are also indicated.

The Chemical Abstracts Service maintains its various general literature and patent indexes in computer-readable form and makes them available for on-line searching through System Development's ORBIT, Lockheed's DIALOG, and other on-line systems. Furthermore, three files are offered covering the periods 1967–1971, 1972–1976, and 1977 to present. The index is searchable on-line at the same time the abstracts appear in the printed edition. Because of arrangements recently made with INPADOC for receiving patent-family data, CAS's patent concordance is no longer available for on-line searching (18). It is replaced as a printed publication by the new patent index mentioned above.

Derwent Publications, Ltd.

Derwent Publications, Ltd., was founded in 1950. Its first publication was a simple abstract bulletin of British patents. Gradually, the number of countries covered was expanded. During the 1960s, chemical coverage was broadened and separate services evolved covering pharmaceutical patents (FARMDOC), agricultural patents (AGDOC), and polymer patents (PLASDOC). These were later (1970) expanded into a full chemical service called the *Central Patents Index* (CPI). The varieties and complexities of the accompanying indexes grew with each expansion. In 1974, the *World Patents Index* (WPI) was created to provide fuller coverage of nonchemical patents (19). Services in this area have been broadened again in 1980 in the form of abstract booklets and indexes covering electrical patents.

Currently, the various Derwent products include about 200,000 basic patents (new to the Derwent data base) each year, along with references to over 200,000 equivalents. A full range of patent-information services is offered, including a number of printed bibliographic and abstract publications, associated printed indexes, on-line searchable files, and a search-bureau operation. The printed publications include the *World Patents Index* (WPI), the *World Patent Abstracts* (WPA), the *Central Patents Index*, and the new *Electrical Patents Index* (EPI). They are organized on the basis of country origin and/or subject content. For the latter purpose, Derwent has developed a scheme that is used at some level of specificity in most of its publications. It classifies chemical technology into 12 sections and nonchemical information into three (see Table 7). The main sections are then further subdivided into a number of classes that arrange citations within certain abstract bulletins and prepare indexes for others.

It is generally agreed that Derwent services offer excellent general-purpose alerting and retrospective-search capabilities. Although they are surpassed by other

Table 7. Main Derwent Classifications

Classification	Code
Chemical	
PLASDOC, polymers	A
FARMDOC, pharmaceuticals	B
AGDOC, agricultural chemicals	C
food, fermentation, detergents, cosmetics, etc	D
CHEMDOC, general chemicals	E
textiles, paper, cellulose	F
printing, coating, photographic chemicals	G
petroleum	H
chemical engineering	J
nuclear, explosives, protection	K
glass, refractories, ceramics, electrochemistry	L
metallurgy	M
Nonchemical	
general (human necessities, performing operations)	P
mechanical (transport, construction, mechanical engineering)	Q
electrical	R

Table 8.   Countries Covered in Derwent Services

Covered fully	Covered by titles and bibliographic data
Belgium	Austria
Canada	Brazil
European (EPO)	Czechoslovakia
France	Denmark
FRG	Finland
GDR	Hungary
Japan[a]	Israel
Netherlands	Italy
PCT (WPO)[b]	Norway
Research Disclosure	Portugal
South Africa	Romania
Sweden	
Switzerland	
United Kingdom	
United States	
USSR	

[a] Chemical patents only.
[b] PCT = Patent Cooperation Treaty; WPO = World Patent Organization.

patent information tools in individual areas (for example, the *Chemical Abstracts* data bases, APIPAT, and IFI's *Comprehensive Database* are superior in various aspects of retrospective searching) no other data base offers such a variety of services (20). However, the Derwent system, in offering so many options, has become highly complex and requires careful attention for the most effective use (21). The countries covered in the various Derwent services are listed in Table 8.

The WPI Gazette.   This is a weekly alerting bulletin providing bibliographic citations and expanded, informative titles for all newly published patent documents. In the *WPI Gazette*, all patents are covered for all countries except Japan, for which only chemically related patents are included. The gazette is published five to six weeks after document publication, in four sections: Ch (Chemical A–M), P (General), Q (Mechanical), and R (Electrical). Each entry provides an informative title and, in a highly stylized format, the critical bibliographic elements: patent number, title (informative, written by Derwent staff), patentee/assignee (not inventor when there is a corporate patentee), priority application and date (earliest and latest only), international patent class, publication date, Derwent accession number, and Derwent class.

The patents announced in each issue are arranged by both patentee (applicant) and international patent class (IPC symbol). In addition, basic-to-equivalent and patent number-to-Derwent accession number indexes are provided. The information presented in the *WPI Gazette* forms the basis for the headings of all Derwent abstracts, though in slightly different format in different publications.

Alerting Abstract Services.   *World Patent Abstracts (WPA).*   This weekly abstract publication includes a series of alerting bulletins for individual countries (see Table 9) and a series of subject-based booklets. The service is published one week after the *WPI Gazette* except for the *European Patent Bulletin*, which appears four to five weeks after the patent document is published, and the *PCT Bulletin*, which appears five to seven weeks after publication.

Table 9. WPA Country Bulletins [a]

Country	Bulletin sections offered
Belgium	chemical; general–mechanical–electrical
EPO	chemical–general–mechanical–electrical
France	chemical
FRG	
Ausleeschriften	chemical; general–mechanical–electrical
Offenlegungsschriften	chemical; general–mechanical; electrical
Japan	
examined	chemical
unexamined	chemical
Netherlands	chemical
United Kingdom	chemical; general–mechanical–electrical
United States	chemical; general–mechanical; electrical
USSR	chemical; general–mechanical; electrical
WPO	chemical–general–mechanical–electrical

[a] WPA = *World Patent Abstracts;* PCT = Patent Cooperation Treaty; WPO = World Patent Organization.

Except for the European and PCT bulletins, which contain lengthy documentation-type abstracts, the abstracts in these publications are relatively short and of the alerting type. They are based generally on the claims, but usually include, in addition, information on the uses or advantages of the invention. Each entry also carries bibliographic data and a title. Since 1975, Derwent has published six subject-based alerting abstract booklets in nonchemical areas. In 1980, with the introduction of the *Electrical Patents Index* this has been expanded to seven booklets (see Table 10). The subject content of each of the booklets in this series is defined in terms of broad Derwent classification; within each booklet the abstracts are further arranged by a more detailed class breakdown.

Table 10. WPA Nonchemical Subject-Based Alerting Booklets [a]

Subject	Derwent class
human necessities	P1–P3
performing operations	P4–P8
transport, construction	Q1–Q4
mechanical engineering	Q5–Q7
instrumentation, computing	S–T
electronic components, circuitry	U–V
communication, electric power	W–X

[a] WPA = *World Patent Abstracts.*

Country-Order Abstract Booklets. As complementary abstract services, Derwent provided in 1980 two series of alerting booklets, arranged by country; one series covers chemical patents and the other electrical patents (EPI, beginning in 1980). They are produced weekly and cover documents published six to seven weeks earlier. The chemical-patent service issues 12 booklets, the EPI service six, one each covering the subject categories shown in Table 11.

These booklets tend to be an attorney's tool, since they are arranged by country,

Table 11. Country-Ordered Abstract Booklets

Code Subject	Code Subject
CPI	
A—PLASDOC (polymers)	G—photographic, coatings, adhesives
B—FARMDOC (pharmaceuticals)	H—petroleum
C—AGDOC (agriculturals)	J—chemical engineering
D—food, cosmetics, water, etc.	K—nuclear, explosives, protection
E—CHEMDOC (general chemicals)	L—glass, ceramics, electrochemistry
F—fibers, textiles, paper	M—metallurgy
EPI	
S—instrumentation, measuring and testing	
T—computing and control	
U—semiconductors and electronic circuitry	
V—electronic components	
W—communications	
X—electric power engineering	

are not subdivided by subject, and repeat information (either full abstract or bibliographic citation) each time an equivalent document is published. Each booklet includes indexes by patentee, Derwent class, accession number (showing family members published to date), and patent number. In addition, a separate index booklet appears each week covering the entire CPI, with indexes by patentee, accession number and patent number.

Classified Alerting Abstract Booklets. A week after the country-ordered booklets are published, abstracts in the 12 CPI sections and six EPI sections are rearranged by their Derwent classes and republished. The classified arrangement, by allowing the scanning of selected subject areas, is geared to the needs of the technical user. In order to reduce redundancy, abstracts are not republished for equivalents except where they have been examined (eg, by the United States, or FRG Auslegeschrift).

Indexes and Retrospective Searching. In addition to the indexes accompanying the various publications noted previously, Derwent prepares cumulative indexes for all of its CPI, WPI, and EPI services. These are arranged by patentee, patent number, accession number, international class, and priority-application number. Various quarterly and annual equivalent concordances are also available. Most cumulative indexes are produced in microfilm rather than as printed publications.

For retrospective retrieval purposes, Derwent has devised a number of detailed coding schemes including manual codes, punch codes, *Ring Index* codes, and PLASDOC key terms. The manual codes are a more detailed breakdown of the previously mentioned class or subsection headings. For example, the code AO4-GO2E2 represents polyethylene applications in film and packaging, whereas the code JO1-KO3 represents flotation or differential sedimentation. There are about 4500 such codes and they are used with the *Basic Abstract Journal*, the manual-code cards, the alerting profile publications, and the on-line files. Punch codes are used for indexing chemical compounds and their uses and, in a separate system, for indexing polymer patents. The coding is restricted to the 960 positions of an 80-column punch card and their paired combinations. Originally, for searching, the punched abstract cards were sorted on a card sorter to produce a set of abstracts for subsequent manual screening. This procedure, however, has given way to computer searching since the same data are now available on line. In the chemical system, compounds and families of compounds are indexed in terms of their structural features, each code or code combination representing a different structural concept. In 1972, *Ring Index* numbers were added to the chemical-indexing scheme and made searchable in certain printed indexes, by computer, and on line. These are not part of the punch-code system. The chemical system (punch codes and *Ring Index* numbers) is used to index the significant compounds found in the FARMDOC, AGDOC, and CHEMDOC (sections B, C, and E) patents. Incidently, specific compounds are not indexed as such in the Derwent scheme nor are other than the end products of chemical reactions. It is understood, however, that among other improvements scheduled for 1981, the chemical punch-code scheme will be replaced with an alpha-numeric code system designed for computer searching and that 2000–3000 common chemicals will be registered for indexing as specific concepts for the first time. Role indicators (functional qualifiers) will also be used. The punch-code system used for polymer patents (PLASDOC, section A) provides for the indexing of polymers in terms of the monomers from which they are made, as well as manufacturing processes, catalysts, additives, and uses. A fault of the punch-code system has been the inherent potential for the false coordination of codes. For example, in polymer patents frequent false retrieval is created because of the false coordination of codes that refer to different polymers indexed for the same patent. To alleviate this problem, the use of a number of precoordinated or bound terms to describe specific polymer concepts (PLASDOC key terms) was initiated in 1978.

Basic Abstract Services.   In addition to the alerting abstracts mentioned above, Derwent provides comprehensive, documentation-type abstracts for its chemical patent services. They cover the basic patents entered into the CPI system. They are available two weeks after the country-alerting abstracts, thus about two months after publication of the patent document. These documentation abstracts contain a general overview of the patent's content, a statement of advantages, a section giving significant detail, and often a specific example. Where applicable a drawing is provided. In general, these are the most detailed and informative abstracts being produced anywhere today.

Basic Abstract Journal.   This publication is offered weekly in 12 issues corresponding to the 12 chemical (CPI) sections. An important feature is the inclusion of an in-depth subject index based on the manual codes. Additional indexes are arranged by accession number, patentee, and patent number. The journal is available in printed or microfilm format.

Profile Booklets.   Because the polymer and general chemistry sections of the CPI services are so large, Derwent produces about 20 profile booklets of basic abstracts selected by their manual codes. Most offerings are from the polymers area, several from general chemistry, one from the petroleum area, and one on catalysis that includes patents from several CPI sections. The profile booklets appear about two months after publication of the patent document.

Manual-Code Cards.   Historically, one of Derwent's most important services in the chemical area has been its manual-code cards. For each section of CPI and forthcoming in EPI, Derwent produces sets of cards containing for CPI, basic abstracts and for EPI, alerting abstracts, grouped by manual codes. The fact that many organizations continue to maintain searchable collections of these cards, in spite of their space and filing time requirements, attests to their value, particularly as a browsing tool. However, these and other manually updated files are becoming increasingly impractical to maintain and within the next few years will surely give way to computer-assisted indexes, especially to on-line interactive services.

On-Line Files.   Essentially all Derwent files are searchable on line through facilities of the System Development Corporation. The earliest information, which extends back to 1963, covers pharmaceutical patents. Full chemical coverage was not developed until 1970. Accessibility to the files and hourly rates are dependent upon the user's current subscription to Derwent's printed services (see Table 12).

Patentees are represented in the Derwent system by four- and five-character codes. Patentee names (19 to 24 characters) became searchable during 1980. However, since some of the early records have only patentee codes, searching by a combination of codes and names is necessary. Inventors have been included in the on-line data base for basic patents since 1978. As noted previously, Derwent classes are broad designations originally created to assign patents to their proper places in Derwent's various printed publications. Although they are not used as primary-search terms for on-line searching, they may be helpful in limiting the output of an otherwise voluminous search. Derwent titles are generally very informative; however, because of the editing that was done in preparing title words for the on-line data-base searching of title words must be carried out with a thorough understanding of the procedures used. International patent classes (IPC) assigned to basic patents and new IPCs assigned to equivalents are entered into the file. However, an additional IPC differing from a class already in the file only in the subgroup is not entered. For example, B01d-059/44 would not be entered if B01d-059/24 was already posted for the patent. With respect to equivalents, each record retrieved in a search shows all known members of a patent family that have the same latest priority. Related patents that share other priorities are also retrievable. Many nonconvention equivalents are identified manually by Derwent and are added to the appropriate patent-family record.

The search parameters mentioned above are available to all Derwent subscribers.

Table 12.   Parameters Searchable in the Derwent On-Line File

Accessible to all users	Accessible to subscribers
accession number	manual code
accession year	punch code
priority application number	*Ring Index* code
priority country	polymer key term
patent number of any basic or equivalent	
patent country of any basic or equivalent	
SDC[a] update code	
patentee (applicant)	
inventor	
Derwent class	
title words	
international patent class	

[a] SDC = System Development Corporation.

Certain additional data elements, however, are accessible only to full subscribers to the Derwent sections involved, or to principal Derwent subscribers, or are accessible at higher rates to other Derwent subscribers. These include manual codes, punch codes, *Ring Index* numbers, and certain polymer terms. Manual codes, specific subgroups of the main Derwent classes, as with the broad Derwent classes mentioned above, are particularly useful in on-line searching for limiting the output of a broad title word or IPC search. If the manual code is to be the main search parameter, a search of the manual-code cards is suggested instead of an on-line search since the volume of answers can be very great and ultimate screening of the abstracts is frequently necessary.

Patent Copies.   Derwent makes available microfilm copies of the complete specifications of all basic documents (as well as U.S. and UK equivalents) covered in the CPI system except Japanese. Microfilm sets of all patents are also available for a number of countries.

IFI/Plenum Data Company

IFI/Plenum Data Company, originally Information for Industry, Incorporated, but now a division of Plenum Publishing Corporation, has provided patent-information services since 1952. During the 1950s, IFI's sole product was the *Uniterm Index to United States Chemical Patents*, comprising a dual-dictionary-type keyword index in book form and a companion volume containing OG claims or abstracts of the accessioned patents. Today IFI produces, in addition to the Uniterm book index, two variations of a magnetic-tape index, a number of printed publications including the *IFI Assignee Index* and the *Patent Intelligence and Technology Report* and several data bases for on-line searching. The latter cover all U.S. patents back to 1963 and U.S. chemical patents to 1950. In addition, IFI provides weekly customized profiles on U.S. patents and a number of other search services. IFI is an agent for INPADOC services in the United States and Canada.

The IFI Patent Gazette Database.   This data base contains bibliographic data and the text of the OG claim or abstract for U.S. patents, reissued patents, and defensive publications. The file covers bibliographic data for all U.S. patents issued since 1963 and includes the OG claim or front-page abstract since 1971. In addition, bibliographic data for chemical patents extend back to 1950 and OG claim text back to 1963. The bibliographic information includes the following data elements: document number, publication date, application number, application date, priority number, priority country, priority date, divisional filing data, continuation filing data, IPC symbols, U.S. classification codes, patent type (mechanical, chemical, electrical), applicant name(s), inventor name(s), title text, abstract text, and linear chemical-structure codes.

The two-dimensional chemical structures, including generic and Markush representations appearing in the claims or abstracts are coded into a linear machine-readable form and accompany the text. Standard chemical-naming conventions are used to create codes whose interpretation requires no special training beyond a good understanding of chemical nomenclature. A sample of the coding is found in Figure 3. The gazette data base is updated weekly on a seven-day turnaround basis. Using a text-string-search system developed by IFI, documents may be retrieved on the basis of any of the bibliographic-data elements, and any of the words of the annotated titles or the abstract text. The search system provides for left- and right-hand truncation

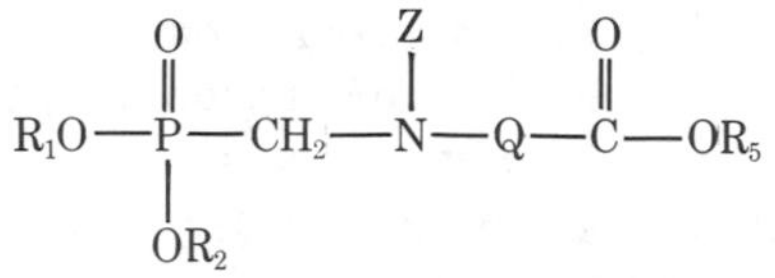

$$R_1O-\overset{\overset{\displaystyle O}{\|}}{\underset{\underset{\displaystyle OR_2}{|}}{P}}-CH_2-\overset{\overset{\displaystyle Z}{|}}{N}-Q-\overset{\overset{\displaystyle O}{\|}}{C}-OR_5$$

R1—O—P(=O)(—O—R2)—CH2—N(—Z)—Q—COO—R5

(a)

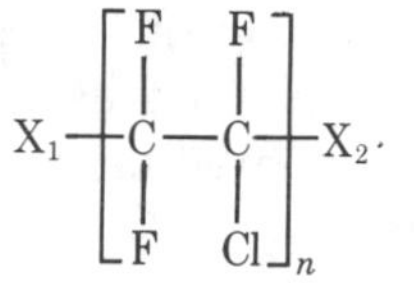

$$X_1\text{---}\left[\begin{array}{c}\overset{\displaystyle F}{\underset{\displaystyle F}{|}}\\ C\\ \end{array}\begin{array}{c}\overset{\displaystyle F}{\underset{\displaystyle Cl}{|}}\\ C\\ \end{array}\right]_n\text{---}X_2\cdot$$

X1—(C(—F)2—C(—F)(—CL))N—X2

(b)

$$R_1\text{---}CH\text{==}\overset{\overset{\displaystyle R_3}{|}}{C}\text{---}NH\text{---}(CH_2)_n\text{---}\underset{\underset{\displaystyle R}{|}}{N}\!\!<\!\!\rceil$$

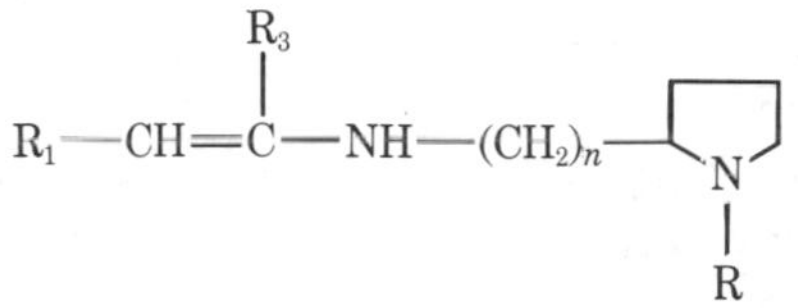

1—R,2—(R1—CH=C(—R3)—NH—(CH2)N—)PYRROLIDINE

(c)

Figure 3. IFI chemical-structure coding.

of any of the search parameters and full use of AND and OR logic. The output can be sorted on the basis of assignee (applicant) name, class code (IPC or U.S.), and ascending or descending document number. It can be selected to provide bibliographic data only, abstract text only, or both bibliographic data and abstract text. The output media can be printed copy or magnetic tape.

Weekly Patent Profiles. Customized weekly alerting reports known as *U.S. Patent Profiles* are available from the *IFI Gazette Database.* The reports are normally mailed seven days after the OG is published. Profiles can be set up, for example, to watch for the issuance of a particular patent, to watch for patents issued to a particular company, to alert a research group to patents issuing in a particular technical area, or to provide an attorney information on new issues pertinent to a pending case.

Patent Gazette Tape File. Another service provided through the *Patent Gazette Database* is a magnetic-tape file. IFI can supply a machine-readable magnetic-tape file of all or part of the data base. For example, a file of polyurethane patents from 1950 to date can be furnished and then updated with ongoing weekly supplements. The standard IFI profile search system is supplied with this tape service.

Customized Intelligence Reports. The *Patent Gazette Database* provides the source material for another patent-information service. IFI custom-designs a patent data analysis, technological forecast, or intelligence report to a user's specifications. The data in this file (1963 to date) can be manipulated and arranged in a number of unique ways. Reports can be developed to help shed light on new areas of technology, on a competitor's activities, or to provide a specialized collection convenient for analysis by an attorney. Examples of special reports include activity profile for a particular company (see under Patent Intelligence and Technology Report); a complete listing of all companies that received a patent in a particular subject area; and a list of all patents issued to a competitor, arranged by U.S. PTO classification and including bibliographic data and patent titles.

On-Line Files. Bibliographic data, claims and/or abstracts and keyword indexing to U.S. patents are maintained for on-line searching through the Lockheed DIALOG Information Retrieval Service. The data bases are known as the CLAIMS files and include the bibliographic data covered in the *Patent Gazette Database* and the keyword indexing covered in the Uniterm index file. *CLAIMS Class*, an index to the U.S. patent classification system is also offered for on-line searching.

Bibliographic Data Bases. Bibliographic data are organized into one current alerting file and three retrospective files based on period covered and content. Although their contents vary slightly, all files contain the following searchable parameters: words in the title (and claim or abstract where available), U.S. PTO classification codes, assignee names, and inventor names. The current file (*File 125 CLAIMS U.S. Patent Abstracts Weekly*) is updated weekly, usually within three weeks of issue. It provides users with the fastest machine-searchable index to U.S. patents available. A sample printout from an on-line bibliographic search is shown in Figure 4.

Key-Word Index Data Bases. Beginning in 1980, IFI provided on-line access to the key-word indexing of their *Uniterm Index of Chemical Patents* (22). This is the same indexing provided in IFI's Uniterm book and tape indexes. The three controlled indexing vocabularies (general terms, chemical terms, and fragment terms) used in producing the index average 35–40 descriptors per patent document. The thesaurus-controlled general-term vocabulary contains about 10,000 descriptors, whereas the registry-controlled compound-term vocabulary contains about 14,000 specific compound names, and the fragmentation system uses about 11,000 fragment concepts. The Uniterm on-line data base is arranged in time periods to correspond with the three retrospective bibliographic files mentioned above; searches of any one of the Uniterm files provide access to the corresponding bibliographic data. A detailed summary of the organization and content of the *CLAIMS* on-line data bases is shown in Table 13.

U.S. PTO Classification Data Base. The *CLAIMS Class* file is an index to the U.S. patent-classification system. Its purpose is to provide users with a means for identifying classes pertinent to a search of the patent-search-center files or other data collections based on U.S. patent classification. The U.S. patent classification system divides subject matter into about 400 main categories called classes. Each class is further divided into subclasses arranged in a hierarchical fashion. There are over 90,000 subclasses in the patent-classification system. Each class and subclass is identified by a unique code number. The subject matter contained in a subclass is indicated by a descriptive title. Lists of subclass numbers and titles showing the hierarchical arrangement within the class are called class schedules. All class schedules are assembled in the *Manual of Classification*. The *CLAIMS Class* data base contains the 90,000+ code numbers and titles of the class schedules. By searching title words the searcher is alerted to the code numbers of classes likely to be useful in searching a primary index based on U.S. patent classification. The file provides a unique access to the *Manual of Classification*.

QUESTION: Find references to the use of diamonds in grinding and abrasive applications.

SEARCH:

```
File25:CLAIMS/US Patent Abstracts - 71-80/Sep
See files 23,24,125
          Set Items Description (+=OR;*=AND;-=NOT)
          --- ----- ------------------------------

? SDIAMOND?
          19    731 DIAMOND?
? SCL=423466000
          20     44 CL=423466000
? SGRIND?
          21   3600 GRIND?
? SABRAS?
          22   3342 ABRAS?
? C(19OR20)AND(21OR22)
          23    166 (19OR20)AND(21OR22)
? T23/5/1-3
23/5/1                   One sample of the 166 answers found.
8380793   8014734
    C/      COMPOSITE  COMPACT  COMPONENTS  FABRICATED  WITH  HIGH TEMPERATURE
BRAZING FILLER METAL AND METHOD FOR MAKING SAME
   Inventor: KNEMEYER FRIEDEL S
   Assignee Names: GENERAL ELECTRIC CO    Assignee Codes: 33808
   Patent No.: 4225322
   Issue Date: 800930
   Application (Date;No): 780110; 868357
   Abstract: A COMPONENT COMPRISED OF A COMPOSITE COMPACT,  PREFERABLY
DIAMOND , AND A SUBSTRATE BONDED TO THE COMPACT.  A PREFERRED EMBODIMENT OF
THE COMPONENT IS A CUTTER FOR A DRILL BIT.  THE COMPACT IS COMPRISED OF  A
LAYER  OF  BONDED  DIAMOND  OR  BORON NITRIDE PARTICLES AND A BASE LAYER OF
CEMENTED CARBIDE BONDED,  PREFERABLY UNDER HIGH TEMPERATURES AND PRESSURES,
TO  THE PARTICULATE LAYER.  THE PARTICULATE LAYER IS DEGRADABLE BY EXPOSURE
TO TEMPERATURES ABOVE A PREDETERMINED TEMPERATURE.  THE SUBSTRATE IS BONDED
TO THE BASE LAYER OF THE COMPACT WITH A FILLER METAL WHICH, TO FORM A BOND,
REQUIRES  THE  EXPOSURE  OF  THE  SURFACES  TO  BE  BONDED TO A TEMPERATURE
SUBSTANTIALLY  GREATER THAN THE DEGRADATION TEMPERATURE OF  THE  PARTICULATE
LAYER. THE COMPONENT IS FABRICATED BY HEATING THE BASE LAYER,  FILLER METAL
AND SUBSTRATE TO A TEMPERATURE IN EXCESS OF  THE  DEGRADATION  TEMPERATURE
WHILE  MAINTAINING  THE  TEMPERATURE  OF  THE  PARTICULATE  LAYER BELOW THE
DEGRADATION TEMPERATURE VIA A HEAT SINK.
      Claim:
      D R A W I N G
      1. AN IMPROVED COMPONENT OF THE TYPE COMPRISED OF:
      A. AN ABRASIVE COMPOSITE COMPACT MADE OF A LAYER OF BONDED
      ABRASIVE PARTICLES SELECTED FROM THE GROUP CONSISTING OF
      DIAMOND AND CUBIC BORON NITRIDE BONDED TOGETHER, SAID
      LAYER BEING BONDED TO A BASE LAYER MADE OF A MATERIAL
      SELECTED FROM THE GROUP CONSISTING OF CEMENTED METAL
      CARBIDE SELECTED FROM THE GROUP CONSISTING OF TUNGSTEN
      CARBIDE, TITANIUM CARBIDE AND TANTALUM CARBIDE WHEREIN
      THE MATERIAL PROVIDING THE METAL BOND IS SELECTED FROM THE
      GROUP CONSISTING OF COBALT, NICKEL, IRON AND MIXTURES
      THEREOF; AN ELEMENTAL METAL WHICH FORMS A STABLE NITRIDE OR
      BORIDE; AND A METAL ALLOY WHICH FORMS A STABLE NITRIDE OR
      BORIDE;
      B. WHICH COMPOSITE COMPACT IS BONDED TO A SUBSTRATE MADE
      OF A MATERIAL SELECTED FROM THE SAME GROUP USED FOR THE
      BASE LAYER;
      C. BY A LAYER OF BRAZING FILLER METAL DISPOSED BETWEEN SAID
      ABRASIVE COMPACT AND SAID SUBSTRATE; WHEREIN, THE IM-
      PROVEMENT COMPRISES THOSE COMPONENTS IN WHICH THE
      BRAZING FILLER METAL HAS A LIQUIDUS SUBSTANTIALLY ABOVE 700°
      C. AND THE THERMAL DEGRADATION TEMPERATURE OF THE LAYER OF
      BONDED ABRASIVE PARTICLES.
   Class: 051295000
   IPC: E21B-009/36
```

SEARCH LOGIC

DIAMOND(S)   or   Class   423-466

and

GRIND····   or   ABRAS····

Figure 4.   IFI on-line search.

IFI Assignee Index. The *IFI Assignee Index* is published quarterly with annual and triannual accumulations. The quarterly editions are published within four weeks of the last patent issue of the quarter and the annual accumulations within ten weeks of the end of the year. Patents listed under each assignee are arranged by U.S. patent classification and include the patent title for each entry.

Table 13.   Data Elements Searchable in IFI *CLAIMS* On-Line Data-Base Files

Data element[a]	Bibliographic abstract files[b], No.				Chemical key-word files[c], No.		
	23	24	25	125	223	224	225
patent numbers	X	X	X	X	X	X	X
title words	X	X	X	X	X	X	X
abstract words			X	X			X
OG claims			X[d]	X			X
assignees	X	X	X	X	X	X	X
inventors	X	X	X	X	X	X	X
OR class codes	X	X	X	X	X	X	X
XR class codes	X	X	X	X	X	X	X
IPC codes			X	X			X
foreign patents	X	X[e]	X[e]		X	X	X
CA references	X	X[e]	X[e]		X	X	X
issue dates		X	X	X		X	X
application information			X	X			X
priority information			X	X			X
year codes	X	X	X	X	X	X	X
patent-type codes (mech, elect, chem)	na[f]	X	X	X	na[f]	na[f]	na[f]
Uniterms					X	X	X
CAS Registry Numbers						X	X

[a] OG = *Official Gazette* (U.S.); OR = original reference; XR = cross-reference; IPC = International Patent Class; CA = *Chemical Abstracts;* CAS = Chemical Abstracts Service.

[b] File 23—Chem 50–62; File 24—U.S. Patents 63–70; File 25—U.S. Patent Abstracts 71+; File 125—*U.S. Patent Abstract Weekly.*

[c] File 223—Uniterm 50–62; File 224—Uniterm 63–70; File 225—Uniterm 71+.

[d] Chemical patents 1971+; all patents 1978+.

[e] Chemical patents only.

[f] Not available.

Patent Intelligence and Technology Report.   The *Patent Intelligence and Technology Report* provides information concerning the ownership and subject distribution of U.S. patents (23–24). As a unique feature, the report includes a six-year analysis of the patent activities by subject area, for each large corporate and institutional assignee. It is designed to be a desk-top tool for corporate managers, research directors, patent-licensing managers, marketing directors, patent attorneys, and financial analysts. The *Patent Intelligence and Technology Report* in combination with the *IFI Assignee Index* enables the user to have a complete picture of all significant U.S. patent activity for the year. The report contains data on all assignees who received 10 or more patents in the year; the report is arranged in four sections.

Alphabetical Listings of Companies.   The alphabetical listing indicates the total number of patents granted during the year, the rank number, and the patent-activity-profile page number for each of the companies in the report.

Listing by Rank.   The rank listing shows each company in descending order according to the number of patents granted.

Distribution of Patents by Company Within U.S. Classification.   Records for all patents processed for the report are arranged by U.S. patent classification (class level only) and then sorted by company. Companies are listed under each class heading in descending order according to the number of patents issued in the particular patent office class.

Patent Activity Profiles.   A six-year patent-activity profile for each company is provided. The profile shows the U.S. patent class number under which the company received patents during the six-year period. The number of patents received in each year is shown in tabular form. Annual totals are shown on the bottom line of the profile. Definitions for the class numbers are found in Section 3.

The Comprehensive Data Base.   During the early 1960s, IFI converted its Uniterm book index to a magnetic-tape data base and provided, on a subscription basis, the tape and a search program for in-house searching (25–26). This early coordinate index was, for the most part, based on the natural language of the patent. Indexing averaged 30 to 35 descriptors per document. Little attempt was made to combine synonyms. This tended to create ambiguity. As the file grew, the lack of indexing-descriptor control lead to increasing false retrieval. In order to alleviate this problem, a weighted-term search program was developed at Gulf Oil (27–29), a subscriber to the file. The program was later purchased by IFI and provided to all subscribers. In spite of this, by 1972 IFI subscribers indicated that there was a need for a more highly controlled indexing system. They suggested that a chemical substructure search capability and a more selective index to polymer chemistry be developed. After reviewing the needs of the industry, IFI decided to purchase the DuPont patent-indexing system which dated back to 1964 (30). The DuPont file incorporated a systematic fragmentation procedure, a compound registry, and a thesaurus-controlled general-term vocabulary. Roles indicated whether the material or compound was a reactant, a product, or just present. Furthermore, a special set of roles was used in indexing polymer patents to distinguish polymers selectively in terms of their monomers (31–32). During 1972, the IFI and DuPont data bases were merged to create the *Comprehensive Database of U.S. Chemical Patents.* This file is searchable from 1950 to date and contains references to over 600,000 patent documents. Searching parameters may be in the form of substructure fragments, compounds, general terms, assignee, or U.S. patent class and subclass codes, including the original and cross references. At the option of the searcher, answer prints can be sorted by assignee, class codes, or by patent number, ascending or descending. Year ranges can also be selected by the searcher. In addition to the patent number, the search printout provides, for each reference, the original patent-classification code, *Chemical Abstracts* reference number, patent title, assignee, and equivalent patents issued in Belgium, France, FRG, UK, and the Netherlands (33). An optional feature incorporated into the system in 1979 allows the user also to print the OG claim or abstract for each answer.

The *Comprehensive Database* is available for in-house searching on an annual subscription basis. The computer-search program is designed for use on the IBM 360/370 series computer and is written in PL-1 language. The master file is updated and forwarded to subscribers quarterly. Technical assistance is provided for loading the file and for training searchers. A sample search printout is shown in Figure 5. In addition to providing the *Comprehensive Database* for searching in-house, IFI searches the file on an individual search-request basis through its search bureau operation at Wilmington, Delaware.

The Uniterm Book and Tape Indexes.   The *Uniterm Index to U.S. Chemical Patents* (book edition) has been published since 1952 and now comprises a 30-year collection, 1950 through 1979. It is updated quarterly and collected annually. Since 1972, controlled vocabularies and computer-edited input have been used for indexing. The index is arranged in dual-dictionary format using a unique numbering scheme to

UNITED STATES CHEMICAL PATENTS

COMPREHENSIVE IFI DATA BASE RETRIEVAL SYSTEM - 1950-1978IFI 12/11/79

SEARCH REQUEST 00006 : SAMPLE SEARCH GE 000

1000 WAS LIMIT ON PRINTED RESULTS RESULTS WERE NOT SAVED ON TAPE

5 DOCUMENTS SATISFIED THIS REQUEST

YEARS SEARCHED : 1950 TO 1979

QUESTION DESCRIPTION:
LIQUID CRYSTAL TRISAZO COMPOUNDS OF THE FORMULA:

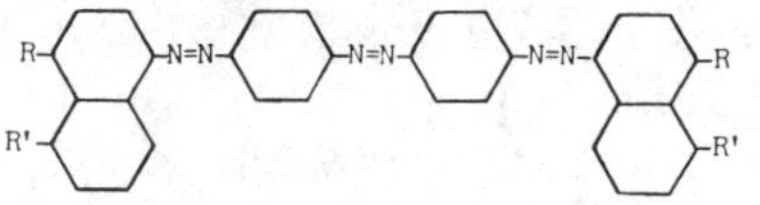

where R and R' are NH_2 or NH-alkyl.

COMPREHENSIVE IFI DATA BASE RETRIEVAL SYSTEM -SEARCH REQUEST 00006 : PAGE 001

ACC. NUMBER	TOTAL WEIGHT	PATENT NUMBER	O.R. CLASS CODE	ASSIGNEE NUMBER	CHEM. AB. REFERENCE	
7121542	105	3,627,699	252 299.000	91840	74081757E	LIQUID CRYSTAL CHOLESTERIC MATERIAL AND SENSITIZING AGENT COMPOSITION AND METHOD FOR DETECTING ELECTROMAGNETIC RADIATION -- WESTINGH--- 000000 DT2019866 GB1309558 FR2040493
7413898	116	3,835,781	101 464.000	68000		/P/METHOD FOR OBTAINING PHOTOGRAPHIC REP RODUCTIONS IN COLOR/IMBIBITION DYE TRANS FER/
7600272	15	3,931,041	252 299.000	93448	84143081X	/P/LIQUID CRYSTAL COMPOSITIONS/IMAGING/ -- XEROX
7819921	117	4,122,027	252 299.000	33808		/C/DICHROIC LIQUID CRYSTAL COMPOSITION W ITH 4,4-BIS (SUBSTITUTED NAPHTHYLAZO)AZO BENZENE DICHROIC DYES/CONTRAST RATIOS — GE
7901703	15	4,137,193	252 299.000	11950		/C/UTILIZATION OF DYES FOR LIQUID CRYSTA L DISPLAYS/AZO DYES -- BROWN SH

Figure 5. IFI *Comprehensive Database* search printout.

simplify coordinate searching. Indexing words are arranged in general terms, compound terms, and fragment terms. The general-term vocabulary includes descriptors for properties, uses, processes, reactions, polymers, natural materials, and trade mixtures. The compound section of the index lists specific structurable chemicals. Inverted word forms have been used frequently in a manner that similar compounds appear together. Separate cross-reference lists are available arranged alphabetically and by molecular formula. Fragment terms are used to index compounds or compound classes not provided for in the general- or compound-term vocabularies. They are particularly useful in the indexing of Markush or generic structures (31). Uniterm also provides indexes based on U.S. classification codes, assignees, and inventors. A tape version of the *Uniterm Index* is available.

The International Patent Documentation Center (INPADOC)

The International Patent Documentation Center (INPADOC) was founded in 1972 following consultations between the World Intellectual Property Organization and the Austrian Government. An agreement was signed on May 2, 1972, and entered into force on June 22, 1973, after ratification by the Austrian Parliament (34).

The industrial property offices of the world collectively issue nearly one million (10^6) patent documents a year. Each document contains bibliographic-data items that can be used to identify, classify, and retrieve the document. The general task of IN-

PADOC is to record, in a computer-readable form, the significant bibliographic-data items of patent documents as soon as they are published and then to manipulate the recorded information to provide information services (35). The services, which are described in detail below, are designed to provide timely and reliable access to patent documents through the recorded bibliographic data.

Collecting and Recording Information.   The following bibliographic-data items are recorded by INPADOC:

Basic items (*1–10*) are recorded for the patent documents published by all countries and organizations covered in the data base. (*1*) Country or organization that published the patent document; (*2*) code to indicate, in respect to the document in question, the kind to which it belongs among the various kinds of patent documents published by that country or organization, eg, patent, inventor's certificate, utility model, unexamined or examined application, etc; (*3*) number of the patent document; (*4*) number of the application; (*5*) filing data of the application; (*6*) publication date of the patent document (or date of an *Official Gazette* announcement); (*7*) international patent classification symbol(s) assigned to the patent document by the publishing country or organization; (*8*) country in or organization with which a priority application was filed; (*9*) application (or filing) number of the priority application in that country or organization; and (*10*) filing date of the priority application.

Additional items (*11 15*) are recorded for countries and organizations that routinely provide such data to INPADOC. (*11*) Name(s) of the inventor(s), (*12*) name(s) of the applicant(s) or owner(s) of the patent, etc, (*13*) title of the invention, (*14*) national classification symbol(s), if any, assigned to the patent document, and (*15*) data concerning other nonpriority applications that have a legal connection with the patent document.

Table 14 lists the countries covered in the INPADOC data base, indicates the data items recorded for each, and shows the earliest date covered. The current data base, starting with the year 1973, includes more than 6.5×10^6 patent documents. INPADOC also maintains a back-file data base consisting of bibliographic-data items for approximately 1.8×10^6 documents issued between 1968 and 1972, inclusive. It includes patent documents published by Austria, Belgium, Canada, France, Israel, Luxembourg, the Netherlands, the Scandinavian countries, South Africa, Spain, Switzerland, the United Kingdom, the United States, the FRG, and Zambia.

INPADOC, with the help of WIPO, has entered into various agreements of cooperation with national industrial property offices and other organizations for the exchange of patent information. As a result, national and regional offices deliver to INPADOC the data for their newly published patent documents. The magnetic tapes containing these data are received by INPADOC at frequent intervals, in most cases weekly. A universal input program then rearranges the data into a standard format to create the INPADOC master data base.

INPADOC Services.   Current prices and supply information may be obtained by writing to IFI/Plenum Data Company (United States or Canada) or INPADOC (elsewhere).

Computer Output on Microfiche (COM).   The main INPADOC services use microfiches as the data carrier. Each microfiche contains 208 frames of computer-generated print with each frame containing approximately 55 lines. Eye-readable headings on the top edge of each microfiche, backed by a white strip, permit identification without a microfilm reader. The last frame on each microfiche is a computer-generated index to the information contained on the other frames.

Table 14. Coverage of INPADOC's Current Data Base [a]

Country or organization	(1–6) (8–10)	(7)	Earliest date of (1–10)	(11)	(12)	(13)	(14)	(15)	Remarks [c]
Argentina	+	+	08.02.73		+	+			
Australia	+	+	18.01.73	+	+	+			
Austria	+	+	10.01.73	+	+	+	+	+	A
Belgium	+	+	02.01.73						B
Brazil	+	+	02.01.73	+	+	+			
Bulgaria	+	+	15.02.73	+	+	+			
Canada	+		01.01.74	+	+	+	+	+	
Cuba	+	+	13.02.74	+	+	+			
Cyprus	+	+	01.03.75	+	+	+			
Czechoslovakia	+	+	22.02.73	+	+	+			
Denmark	+	+	02.01.73	+	+				C [d]
Egypt	+	+	31.01.76	+	+	+			
European Patent Office	+	+	20.12.78	+	+	+			C
Finland	+	+	31.01.73	+	+			+	C [d]
France	+	+	05.01.73		+	+			C, D
German Democratic Republic	+	+	05.01.73	+	+	+			
Germany (Federal Republic of)	+	+	04.01.73	+	+	+			C, E
Greece	+		04.07.77	+	+	+			F
Hong Kong	+	+	05.03.76	+	+	+			
Hungary	+	+	29.01.73	+	+	+			
India	+	+	02.08.75	+	+	+			
International Bureau of WIPO	+	+	19.10.78	+		+			
Ireland	+	+	10.01.73		+	+			
Israel	+	+	30.01.73		+	+			G
Italy	+	+	25.01.73	+	+	+			B
Japan	+	+	02.04.73	+	+	+			C, H, I
Kenya	+	+	11.07.75	+	+	+			
Korea, Republic of	+		08.01.78						
Luxembourg	+	+	05.01.73						B
Malawi	+	+	09.05.73		+	+			B
Monaco	+	+	10.10.75	+	+	+			B
Mongolia	+	+	20.11.72						
Netherlands	+	+	02.01.73		+	+			C
Norway	+	+	02.01.73	+	+			+	C [d]
Philippines	+	+	03.07.75	+	+	+			
Poland	+	+	28.02.73	+	+	+			
Portugl	+	+	01.01.76	+	+	+			
Romania	+	+	20.01.73	+	+	+			
South Africa	+	+	31.01.73	+	+	+			B
Soviet Union	+	+	08.01.73	+	+	+			F
Spain	+	+	01.01.73		+	+			
Sweden	+	+	08.01.73	+	+	+			C [d]
Switzerland	+	+	15.01.73		+	+			
Turkey	+		01.01.73		+	+			
United Kingdom	+	+	04.01.73		+	+			
United States	+	+	02.01.73	+	+	+	+	+	J

Table 14 (*continued*)

Country or organization	(1–6) (8–10)	(7)	Earliest date of (1–10)	(11)	(12)	(13)	(14)	(15)	Remarks[c]
	Bibliographic data items[b]								
Yugoslavia	+	+	28.02.73	+	+	+			K
Zambia	+	+	22.01.73		+	+			B

[a] In January 1979, the INPADOC Data Base contained more than 6.5×10^6 patent documents.
[b] The numbers in italics refer to the bibliographic data items listed under Collecting and Recording Information.
[c] A, Including *Aufgebote* from January 1, 1975, onwards. B, International Patent Class (IPC) symbols to subclass level only. C, Including applications laid open for public inspection before and after examination as well as grants. D, Title and inventor for first publication stage only. E, Including *Gebrauchsmuster* from January 1, 1975, onwards, F. Titles in English from July 1, 1977, onwards. G, Including "applications filed." H, IPC symbols from January 1, 1975, onwards. I, Applicant, patentee, and inventor for the first publication stage only; titles are transliterated from Katakama characters. J, Including "reissues" from July 1, 1975, onwards. K, Titles in English from January 1, 1977, onwards.
[d] From January 1, 1975, onwards.

Patent-Family Service/INPADOC Numerical List (PFS/INL). This service permits the retrieval of all patent documents issued by different countries or organizations that are based upon the same priority application; thus, it identifies patent families. Priority applications are listed first, according to the countries in which they were filed and within each country by priority date and number (PFS part of the service), and second, in the numerical order of all the numbers (irrespective of the country in which they were first filed). Each number is, however, accompanied by an indication of the country in which the priority application was filed (INL part of the service). For each method of listing, patent-family members, ie, the patent documents referring to the same priority application, are then listed in chronological order according to their date of publication in the different countries. Document number, publication date, a code indicating the kind of document, filing date and number of the application, IPC symbol(s), name of the applicant, and title of the invention are given for each patent document. A sample printout of the PFS is given in Figure 6.

Patent-Classification Service (PCS). This service arranges patent documents by IPC symbol and thus groups together documents that relate to similar technological fields. Each patent is listed as many times as there are IPC symbols allotted to it. Each citation carries with it the important additional bibliographic data.

Patent-Applicant Services (PAS). This service groups documents with the same applicant or owner and within each applicant sorts them according to their IPC symbols (main class). Therefore, patents on similar subjects are grouped together under each applicant. For each patent document listed the significant bibliographic data are given.

Patent-Inventor Service (PIS). This service is similar to the patent-applicant service but lists documents according to the name of the inventor. For each inventor the documents are grouped according to their IPC symbols (main class). Thus, it is possible to recognize and retrieve all patent documents attributed to the same inventor or inventors, regardless of the fact that the application was filed by, or the patent granted to, different applicants or owners in different countries, or that priority was or was not claimed.

P F S PATENT FAMILY SERVICE MICROFICHE JAN/1978 73-01-01 / 77-12-31 PRODUCED: 78-01-08 PAGE:148 144
FRAME: L09

CC	PR.**DAT	KP	PRIORITY NO.	CC	PUB**DAT	KD	DOC.NO	APP**DAT	KA	YY	APPL.NO	I P C	APPLICANT	TITLE
SE SE	75-03-18			IN	76-11-20	A	140518	76-03-12	A	76CA	444	G09B 1/00	SVENSKA AB LAMI NATOR	A DEVICE FOR RECEIVING TEMPORA RY MARKINGS
SE SE				ZA	77-02-23	A	761430	76-03-09	A	76	1430	B43L	SVENSKA AB LAMI NATOR	A DEVICE FOR RECEIVING TEMPORA RY MARKINGS
SE SE SE				SE	77-09-05	B	396039	75-03-18	A	75	7503092	B43L 1/12	* SVENSKA AB LA MINATOR	ANORDNING FOR TEMPORERA MARKER INGAR, SERSKILT FOR UNDERVISN INGSENDAMAL
SE SE				AU	77-09-15	A1	11976/76	76-03-12	A	76	11976	G09B 1/04 G09B 19/00	SVENSKA AB LAMI NATOR	TEMPORARY MARKINGS RECEIVER
SE SE SE		A	75 7503097	BR	76-09-14	A	7601574	76-03-16	A	76	7601574	B21J 7/28	ATLAS COPCO AB	DISPOSITIVO AMORTECEDOR DE REC HACA PARA UMA FERRAMENTA DE P ERCUSSAO
SE				FI	76-09-19	A	760664	76-03-12	A	76	760664	B25D	ATLAS COPCO AB	
SE SE SE SE				SE	76-09-19	A	7503097	75-03-18	A	75	7503097	B25D 17/24	ATLAS COPCO AB	ANORDNING VID BERGBORRMASKINER FOR DEMPNING AV REFLEXER FRA N ETT TILL MASKINEN ANSLUTET ARBETSVERKTYG
SE SE SE SE				DE	76-09-23	A1	2610619	76-03-13	A	76	2610619	E21C 15/00 E21C 3/00	ATLAS COPCO AB, NACKA (SCHWED EN)	VORRICHTUNG ZUR DAEMPFUNG DES RUECKSCHLAGS DES WERKZEUGEINS ATZES EINES SCHLAGWERKZEUGS M IT KRAFTANTRIEB
SE SE				JP	76-10-13	A2	51116101	76-03-18	A	76	28694	B25D 9/12 E21C 3/20	ATLAS COPCO AB	SHOGEKIKOGUNOFUKUZAKANSHOSOCHI
SE SE				FR	76-10-15	A1	2304449	76-03-17	A	76	7607667	B25D 17/24	ATLAS COPCO AB	AMORTISSEUR DE REBOND POUR OUT IL A PERCUSSION
SE SE SE				ZA	77-02-23	A	761349	76-03-05	A	76	1349	B25D	ATLAS COPCO AB	DEVICE FOR DAMPING THE RECOIL OF A WORK TOOL CONNECTED TO A PERCUSSION TOOL
SE SE SE SE				SE	77-04-25	B	392830	75-03-18	A	75	7503097	B25D 17/24	ATLAS COPCO AB	ANORDNING VID BERGBORRMASKINER FOR DEMPNING AV REKYLEN FRAN ETT TILL MASKINEN ANSLUTET A RBETSVERKTYG
SE SE SE				AU	77-09-22	A1	12083/76	76-03-17	A	76	12083	E21C 1/02 B25D 17/24 B25D 9/08	ATLAS COPCO AKT IEBOLAG	DEVICE FOR DAMPING THE RECOIL OF A WORK TOOL CONNECTED TO A PERCUSSION TOOL
SE SE SE SE				AT	77-12-15	A	1866/76	76 03-15	A	76	1866	E21C 19/00	ATLAS COPCO AKT IEBOLAG	VORRICHTUNG ZUR DAMPFUNG DES R UCKSCHLAGS DES WERKZEUGEINSAT ZES EINES SCHLAGWERKZEUGS MIT KRAFTANTRIEB
SE	75-03-19	A	75 7503103	DK	76-09-20	A	1163/76	76-03-17	A	76	1163	F16B	ERICSSON L M AB	EKSPANSIONSBOLT
SE SE				FI	76-09-20	A	760371	76-02-16	A	76	760371	F16B	OY L M ERICSSON AB	
SE SE SE				NO	76-09-21	A	760957	76-03-18	A	76	760957	F16L	TELEFONAKTIEBOL AGET L M ERICS SON	
SE SE SE				DE	76-09-30	A1	2607451	76-02-24	A	76	2607451	F16B 35/04	TELEFONAKTIEBOL AGET LM ERICSS ON, STOCKHOLM	SPREIZBOLZEN
SE SE				FR	76-10-15	A1	2304813	76-03-18	A	76	7607847	F16B 13/06	ERICSSON TELEFO NAKTIEBOLAGET	BOULON EXPANSIBLE
SE				JP	76-10-15	A2	51117265	76-03-18	A	76	29791	F16B 19/10	ERICSSON TELEFO	KAKUCHOBORUTO

(C) 1978 INPADOC MICROFICHE PRODUCED BY KODAK AUSTRIA

Figure 6. INPADOC PFS microfiche.

Numerical Database Service (NDB). This service lists patent documents according to the issuing country or organization and the document number. With this service, it is possible to trace all documents that have been published at various procedural steps after the filing of an application (eg, Offenlegungsschrift, Auslegeschrift, or Patentschrift in the Federal Republic of Germany). For each of the patent documents listed all critical bibliographic data are given.

Subscribers to one or more of the above five services receive periodic deliveries of microfiche sets. A set of approximately 200 microfiches for each service is delivered every three months (January, April, July, and October). The information in each quarterly delivery is updated in such a way as to include the information of the preceding quarterly delivery or deliveries of that year. At the end of each calendar year, the information is updated in the same way, so as to include the information of the preceding year or years, up to five years. After five years, a new five-year period of updating and accumulation begins.

A second group of services using COM microfiches is offered by INPADOC under the name of *INPADOC Patent Gazette* (IPG). The IPG gives, once a week, in one set of microfiches bibliographic information relating to newly published patent documents (eg, since the last preceding issue of the IPG) in the countries or by the organizations covered in the INPADOC services (see Table 14). The IPG is thus a collation of information otherwise given only in the patent gazettes of these countries or organizations. The four services that form the IPG are the Selected Classification Service (SCS), the Selected Applicant Service (SAS), the Selected Inventor Service (SIS), and the Selected Numerical Service (SNS). These four services correspond respectively to the PCS, the PAS, the PIS, and the NDB described above. They are offered individually or in combination. In addition to providing significant bibliographic data, the IPG services provide all known patent-family members for each patent document listed. These services are not cumulative.

Weekly Magnetic-Tape Services. *Extended Data Tape (EDT).* This tape contains the basic bibliographic-data items, where recorded, the names of the applicant or owner, and of the inventor, and the title of the invention of those patent documents added to the INPADOC data base during the previous week.

INPADOC Family Data Tape (IFD). This tape service is similar to the Extended Data Tape but includes additional information concerning patent-family members already recorded in the INPADOC data base.

Individual Requests for Information. *Individual Request for Patent Family (IRF).* This service lists patent documents based on the same priority applications. The listed patent documents thus form a patent family.

Individual Request for Patent Classification (IRC). This service provides a list of all patent documents to which a specified IPC symbol has been allotted.

Individual Request for Applicant (IRA). This service provides a list of all patent documents recorded under the name of a specified applicant or owner.

Individual Request for Inventor (IRI). This service provides a list of all patent documents recorded under the name of a specified inventor.

Patent-Document Copy Service. INPADOC maintains one of the largest collections of 16-mm-roll microfilm of patent documents in the world, consisting of more than 30,000 rolls each containing between 200 and 300 documents. Filming is continuous to keep the collection up to date as new documents are issued. The collection includes patent documents issued by the following countries and organizations (the year of publication of the earliest patent document available in parenthesis): Australia (1926), Austria (1899), Belgium (1950), Canada (1950), Czechoslovakia (1919), Denmark (1895), European Patent Office (EPO) (1978), Finland (1954), France (1902), German Democratic Republic (1951), German Reich (1877), Federal Republic of Germany (1945), Hungary (1896), International Bureau of WIPO (1978), Italy (1926), Netherlands (1913), Norway (1892), Poland (1924), Sweden (1885), Switzerland (1889), United Kingdom (1900), United States (1935), Yugoslavia (1922). Copies of patent documents can be obtained from INPADOC either in the form of paper copies of individual documents or in the form of 16-mm-roll microfilm for continuous numerical series of documents.

CAPRI System for the Extension of IPC Symbols. The international patent classification (IPC) has been in general use in most patent-document issuing offices only since about 1970. Consequently, most of the approximately 27×10^6 patent documents issued to date throughout the world do not bear the IPC symbols. Industrial offices are increasingly changing from their domestic classification system to the IPC for organizing their examination and public search files. WIPO and INPADOC signed, in 1975, an agreement concerning the computerized administration of patent documents reclassified according to the IPC (CAPRI System). The aim of the CAPRI system is to collect and store the IPC symbols allotted to patent documents issued before 1975, with priority being given to the coverage, at least initially, of those patent documents issued from 1920 onward by France, the Federal Republic of Germany and the former Reichspatentamt of Germany, Japan, the USSR, Switzerland, the United Kingdom, and the United States.

To implement the CAPRI System, INPADOC currently receives IPC information from Austria, Federal Republic of Germany, the USSR, and the European Patent Office. Cooperation with the Japanese Patent Office is anticipated in the near future. The CAPRI Central Database (CDB) is organized by IPC subclasses and at present contains information relating to 7.5×10^6 documents. The file is expected to be complete in 1982.

Patent Register Service (PRS). In late 1979, INPADOC initiated a new service pertaining to patent-document status. A data base is being built that includes information on various changes recorded from time to time on the legal status of patent documents already in the INPADOC file. The initial input has been from Austria, Switzerland, FRG, the European Patent Organization, France, the Netherlands, and the World Patent Organization. The data recorded include such items as opposition status, lapses, maintenance payments, invalidations, withdrawals, etc. This file should be of particular interest to corporate patent managers and attorneys.

On-Line Services. INPADOC offers the Patent Gazette, the Patent Family Service, and the Patent Register Service for on-line searching.

Early in 1980, the INPADOC IPG information was made available on line through the Lockheed DIALOG system (36). The file includes the most recent six weeks of data received by INPADOC and is updated biweekly. References to approximately 16,000 patent documents per week from 47 countries and two regional offices go into this data base. This file, as well as the Patent Family Service, is also available in Europe through Fachinformationszentrum Energie, Physik, Mathematik GmbH, or directly through INPADOC. The Patent Register Service is offered only through the INPADOC facilities. It is expected that the Patent Family Service will become available in the United States via the Lockheed DIALOG service during 1981.

Other Secondary Patent-Information Sources

Research Disclosure. The research-disclosure service, although actually a primary information source, is discussed here for convenience. As noted above, at times it is economically unattractive to pursue the filing of a patent application or to continue the prosecution of one already filed. In these cases it is desirable to publish a description of the potential invention, for defensive purposes, through an alternative means. The U.S. Patent Office provides the defensive publication for this purpose. A similar service is offered through the monthly journal *Research Disclosure*. This publication was started by Industrial Opportunities, Ltd. in 1960. For a modest fee, a description or disclosure of a potential invention is published and distributed to some 125 patent offices and libraries around the world. Thus, the technology involved becomes available as a prior-art disclosure that could be cited against a subsequent patent application by someone else on the same subject. Descriptions are published within six weeks of receipt by Industrial Opportunities. About 750 to 1000 disclosures are recorded annually. Derwent and Chemical Abstracts provide an abstracts service of material appearing in *Research Disclosure* and IFI/Plenum publishes a key-word index to the subject content.

Patent Abstracts of Japan. In order to promote the dissemination of Japanese technology throughout the world and to provide a national source of information on Japanese patent documents particularly useful to the Western nations, the Japanese Patent Office in 1979 began sponsoring the publication of the English-language abstract gazette *Patent Abstracts of Japan*. The abstracts are prepared by the Japan Patent Information Center (JAPATIC), and the gazette is published and marketed by the Japan Institute of Invention and Innovation (JIII). Included are abstracts for all unexamined patent applications laid open to public inspection but excluding those filed by foreign applicants, private Japanese nationals and applications on technology peculiar to Japan. It is estimated that of the approximately 150,000 applications laid open annually, 70,000 are included in the gazette. The gazette is divided into mechanical, electrical, and chemical sections. Each issue includes 500 abstracts, three to a page. The lag time between the publication date of the application and the publication in the gazette is expected to be two months after an initial start-up period. As of March 1980, abstracts for 250,000 applications published since 1977 have become available (37).

Both JIII and JAPATIC offer additional patent services. JIII was established in 1904 as the Association for the Protection of Industrial Property. Since then it has undergone a number of organizational and name changes and in 1947 acquired its present status. JIII provides, for a fee, subject matter, equivalent, applicant and inventor name, and design and utility-model searches. It also provides English translations of Japanese patent documents. JAPATIC was established in 1971. Except for the English-language abstracts mentioned above and English-language abstracts provided for the International Atomic Energy Agency, it provides services intended principally for Japanese readers. These services are based on computer-assisted systems and magnetic-tape data bases. JAPATIC is the sales agency for INPADOC services in Japan.

Rapid Patent International.   Rapid Patent International is primarily a patent-document-copy supplier. Orders for copies of U.S. patents are accepted by individual number or on a subscription basis by class or assignee. The latter is a unique service and, when coupled with a class subscription, can provide an excellent alerting service. The copy is of high quality, and orders are filled promptly with a minimum of missing or incomplete documents. Copies of foreign patent documents are also available. Rapid Patent offers a watch service and a document filing service, in addition to patent copies, subject searches, translations, assignment searches, and file-history copies. Rapid Patent's offices are adjacent the Patent Office at Arlington, Virginia.

Search Check, Inc.   Search Check, Inc. maintains a computer-readable data base of U.S. patent numbers back to 1947, each having associated therewith both the numbers of the patent references cited against it during prosecution and the numbers of the patents in which it has subsequently been cited. With the help of this data base, it is possible to search both forward and backward from a known reference or key patent to find cited references. In cases where a particularly pertinent patent is known or has been uncovered in the course of a subject search elsewhere, a check of the Search Check file may reveal additional pertinent references. Patents found can, of course, be made the key patents for additional Search Check queries. This unique search technique is particularly useful when an exhaustive search is required. The Search Check data base is expected to be offered for on-line searching in late 1980. It will be marketed by IFI/Plenum as the *CLAIMS Citation* file.

National Technical Information Service (NTIS).   NTIS is a service of the U.S. Department of Commerce. It is the sales agency for government-generated technical information and provides a number of patent-related publications and references. In a unique publication, it provides abstracts of U.S. patent documents resulting from certain government-sponsored research. *Government Inventions for Licensing* is published each Tuesday and contains abstracts for 60 to 70 applications and patents the technology of which is available for licensing from the Federal government. This information is also available through the NTIS on-line data base.

TNO.   TNO is the central organization for applied scientific research in the Netherlands. It was formerly known as NIDER and has had a long history of providing quality watch services and custom searching.

Nippon Gijutsu Boeki Co., Ltd. (NGB).   NGB is a patent-information organization providing a number of services dealing primarily with Japanese patents. They also market Derwent services in Japan. In addition to the usual patent-copy, translation, search, and watch services, NGB offers claims sheets of Japanese patents. These contain bibliographic data, the claim(s) (in Japanese), and one drawing. They are provided in punched-card format or on microfilm.

Polyresearch Service.   Polyresearch Service was founded in 1955. Its main offices are located in Rijswijk, the Netherlands, with branch offices in Canada, the FRG, Italy, and Japan. Polyresearch offers quality patent-information services. Because it is located near the Dutch Patent Office and the examining division of the European Patent Office, Polyresearch has direct access to the search facilities of these organizations. File histories of applications filed in the Netherlands, the FRG, and the United Kingdom are available. Translations of general technical and scientific literature as well as patents are provided. The languages covered include English, German, French, Dutch, Spanish, Italian, Danish, Swedish, Norwegian, Russian, Czech, Yugoslav, and Japanese. In addition, Polyresearch offers the usual complement of other patent-information services.

American Petroleum Institute (API).   The central abstracting and indexing service of the American Petroleum Institute provides comprehensive patent-alerting and indexing services for the petroleum industry. Under an arrangement with Derwent Publications, API uses Derwent abstracts as the source documents for its publications and indexes. Selected Derwent alerting abstracts are rearranged into subject categories of interest to the petroleum industry; the material is then reprinted in a series of weekly subject-oriented alerting bulletins. The bulletins include a generous selection of patents on general chemicals and polymers, in addition to those on petroleum technology. Printed indexes are provided for a subset of the patents that appear in the alerting bulletins selected more strictly on the basis of petroleum interests. These are prepared by reference to the corresponding Derwent documentation abstracts. About 7000 patents are indexed each year. The API patent information is available for on-line searching through the SDC ORBIT system as APIPAT (38–39).

The Rubber and Plastics Research Association of Great Britain (RAPRA).   The Rubber and Plastics Research Association of Great Britain publishes *Plastics RAPRA Abstracts* and *Rubber RAPRA Abstracts* which include extensive coverage of UK, French, FRG, USSR, and U.S. patents. A wide spectrum of rubber and plastics technology is provided including polymerization processes, natural-rubber cultivation, monomers and compounding ingredients and their syntheses, applications, mechanical processing, machinery, and testing and test equipment. The bulletins are published biweekly and the abstracts are arranged by subject matter. Indexes are included in each issue and annual accumulations are provided. The RAPRA information is also available for on-line searching.

Research Publications, Inc.   Research Publications provides two services satisfying special patent-information needs. They produce on microfilm soon after issue the full text of all U.S. patent documents. Since 1973, they have also provided microfilm listings of changes made to issued U.S. patents. The service, known as the CDR File, includes all changes in status listed in the Official Gazette, eg, adverse decisions, dedications, disclaimers, reissues, patents withdrawn, and corrections. The OG citations are provided on microfilm. A printed index refers the user from patent number to microfilm reel number and gives a code indicating the type of change involved. The index, issued annually, is cumulative from 1973.

J. Gevers et Cie.   J. Gevers publishes a monthly alerting service for Belgium patents entitled *Revue Gevers des brevets et répertoire des brevets belges récents*. It lists all applications in IPC order and provides an alphabetical listing by applicant name. Applications appear in this service about five months after filing.

Searching

Over the past decade, patent information has grown at an extraordinary rate, and searching techniques have become more and more diverse and sophisticated. To make matters even worse, recent changes in the patent laws of several major countries "have resulted in sizable increases in numbers of publication and changes in numbering procedures . . ." (9). The individual chemist or an associate in the marketing, business, or legal area is no longer able to keep abreast without some help from the information specialist (40), ie, the expert in the diverse files and data bases that record and store the information and in the advanced procedures and systems that have been developed for searching it. Depending upon the circumstance, the information specialist may simply steer the inquirer in the right direction, or may actually carry out the search, analyze the results, and prepare a written report. The information specialist may be an information-service assistant in the technical library or information-service organization of the inquirer's own company, a service representative of a commercial information supplier, an aid at the patent-office search room, or a librarian at an academic or public library. The inquirer would be well advised to seek the assistance of an information specialist before committing a great deal of time or money to a search, particularly one in an unfamiliar area.

When setting up a search strategy for a particular question, it must be decided whether to include foreign patents and, if so, from which countries. If the search pertains to current awareness, that is, if a prime requirement is early identification of the first publication, inclusion of foreign patents is critical. Even if there is a U.S. counterpart the foreign equivalent is often published significantly earlier. Since infringement can occur only with respect to patents granted by the country in which the potentially infringing manufacture, sale, or use takes place, infringement questions usually involve only one country. For a U.S. company not anticipating manufacture or sales abroad, only U.S. patents would be of concern. In high risk situations, however, even here it may be useful to search foreign patents. Leads to potentially pertinent U.S. equivalents that could issue later might be uncovered. If early identification or infringement is not a factor, the decision on whether to search foreign documents becomes a matter of economics. Currently, approximately 60,000 U.S. patents are issued annually. The publication rate, worldwide, is estimated at nearly 10^6, with about half of these being basics, that is, the first publication of a member of a patent family.

Any extension of the search into foreign patents, even to include just a few of the main countries, can substantially increase the effort required. This is not to say that foreign patents should not be searched; indeed, many times they must be (1,9,41). What it does mean, however, is that the most efficient means available for searching foreign documents needs to be found.

There are many reasons and therefore many approaches to searching the patent literature. The question of thoroughness or recall (42), although difficult, must be considered. Sometimes a few quick references to document a point or provide directions for the preparation of a compound might be all that is required. Recall would then be not important, and the search would be terminated as soon as answers to the problems in question were found. On other occasions, however, it might be necessary to find all pertinent answers. If the inquirer's purpose were to uncover potential infringement by a new manufacturing process or product or to find references that would invalidate a competitor's patent, 100% recall would be desired. Each search, big or small, has a price tag which must be determined before the search is undertaken. Searches can cost between $5 and $5000. Many search tools are available; selecting the right tool for the particular job at hand is half the battle. It is imperative to be aware of the search services and facilities available both internally, through company services, and externally, through national and regional patent offices, government sponsored agencies, learned societies, and commercial services. With the great proliferation of a variety of scientific and technical literature, and particularly of patents in recent years, many costly and outmoded in-house patent files and services are giving way to external sources where many users share the cost. Since the patent document is both specialized and of stylized form, the uninitiated would be wise to spend some time beforehand becoming familiar with its style and arrangement (43).

State-of-the-Art Searches.   This term is generally applied to searches designed to provide the inquirer with general background, frequently in a rather broad subject area. The patent search is usually coupled with a search of the general literature. For instance, if a person has just been assigned to a new research project or given a new marketing area or become involved in preparing a patent application in a new field, there is a need to acquire background and to be brought up to date. Examples of search topics would include halogenation processes, polyurethane chemistry, color photography, and other equally broad subject categories. Normally, the inquirer is looking for a manageable collection of references, each of which provides meaningful information on the subject. The retrieval of every document or of sketchy or trivial information is neither necessary nor desired. The search needs to be substantive but not exhaustive. The on-line files of Chemical Abstracts, Derwent, and IFI/Plenum Data Company are good places to start (33). They will quickly provide a basic collection of references for further study. After analysis of the on-line answers, more extensive searching may be in order, depending upon the results of the preliminary survey. If necessary, the search may then be extended to the classified files at the public-search room of the PTO or to specialized files such as those of the API, the British Rubber and Plastics Research Association (RAPRA) or one of the specialized Derwent files.

Current Awareness.   Once a background of information in an area of interest has been acquired, it is important to keep abreast of new developments. Current-awareness programs and services are designed to fulfill this need.

Traditionally, the prime source of information for current awareness has been the patent gazette. If the needed gazettes are available, this source is still a useful means for providing information on newly issued patent documents. However, the growing availability of timely patent information through secondary sources has increased their use. As of March 1976, 14 on-line data bases containing at least 5% patent information have been reported (13). A recent important addition to this collection has been the *INPADOC International Gazette Database* where bibliographic-search information about newly issued patent documents from 47 countries and two regional offices is available within a few weeks of publication. An up-to-date list of the principal on-line sources of patent information is given in Table 5. Several hard-copy services are also available for current awareness and alerting needs. Derwent provides alerting information in its WPI Gazette, weekly abstract publications, and profile booklets. Most patents covered in the WPI Gazette appear within five to six weeks of their publication. The alerting and basic abstract publications carry the same patents about one to two weeks later (44). The profiles follow one week after the basic abstract booklets. IFI/Plenum Data Company offers weekly alerting profiles of U.S. patents giving full bibliographic data, title, and OG claim or abstract. These are available 7 to 10 days after the patent issues. A microfiche edition of the *INPADOC International Gazette* reports patent documents from most countries in this service about two to three weeks following publication. Patents are usually reported too late in Chemical Abstracts for alerting purposes.

Novelty Searches. Before filing an application for a patent, it is necessary to appraise the novelty of the potential invention. The search intended to uncover possible anticipatory art that helps the patent counsel evaluate novelty is variously called an anticipation, novelty, patentability, or prior-art search. Anticipation has been defined (45) " . . . disclosure in the prior art of a thing substantially identical with the art or instrument for which the patent is sought." The strategy used for a novelty search is similar to that used for a state-of-the-art search but is generally more specific. An alleged invention usually presents a specific target. It is frequently an improvement or modification to an already known process, composition, use, or device. For example, rather than encompassing the whole area of halogenation, an invention is more likely to be directed at a specific ramification, such as high pressure bromination under certain conditions. The search, nevertheless, should not be set up too narrowly, or related and background art may not be found. In the case of high pressure bromination, the search should uncover high temperature chlorination as well. The extent of this broadening to cover functional equivalents depends on the closeness of the particular art involved. A direct anticipatory reference is seldom uncovered. More often, a number of close references are found. Sometimes separate references, when combined, describe the potential invention. Such references can be anticipatory. The patent counsel working up the application needs to know this type of information. If the search uncovers nothing considered pertinent, the closest references should be reported since related information contributes to the preparation of a strong patent application. Novelty searches are normally not as comprehensive as some of the other subject-matter inquiries, because the stakes are usually much higher when validity or infringement is involved. For economic reasons, a patentability search is rarely exhaustive (46).

Infringement Searches. U.S. patent law provides an inventor or assignee the right to exclude all others from practicing the invention for a period of 17 years from the date of issue. Therefore it must be determined if unexpired patents have claims that might be infringed by a contemplated new commercial venture. The search conducted to answer this question is called an infringement or domination search. The decision to apply for a patent frequently triggers an infringement search. The potential new inventor wishes to know whether or not there will be freedom to practice the invention when and if a patent is granted. This is not the only occasion calling for such action, however. An infringement search should be considered any time a new commercial process, product, use, apparatus, or device is being contemplated. Even modifications or changes in existing operations should be considered. A change in a process solvent or even a change in process conditions such as temperature or pressure may require study of infringement possibility. An addition (but not the removal) of an ingredient to a composition already on the market certainly calls for domination clearance.

Because infringement can occur only with respect to patents granted by the country in which the venture in question is scheduled to take place, infringement searches are normally limited to that country's patents. However, under certain high risk situations, selected foreign patents should also be considered. Unlike novelty searching, where there is usually no legal significance to the time period covered, infringement is concerned only with patents that are still in force. A further difference between the two types of searching is the fact that, for infringement purposes, only the claims need to be studied.

In setting up the strategy for an infringement search, each potentially infringing element of the process, composition, apparatus, or device in question must be considered. Again, there is a difference between novelty and infringement searching. In the former, a reference broader than the subject is, at best, only of lesser interest. A patent claiming halogenation in general, with specific examples only to chlorination at room temperature, would not be a good reference in a search investigating the novelty of a high temperature bromination process. However, in an infringement search, such a patent could be pertinent. If the broad claim were general to halogenation and contained no other limiting elements not also present in the search process, the patent would require legal study to determine whether or not it might be judged infringed. In a sense, the logic of an infringement search is the opposite of that in a novelty search. Anticipatory references are selected on the basis of certain information being present. Potentially dominating patents are the ones that cannot be rejected on the basis of restricting or limiting elements of the claims. In the former case, selection is by positive action; in the latter, by negative action. As expressed in *Inventions and Their Management* (47), "A claim does not dominate an invention (or commercial venture) that includes less than every element or part mentioned in the claim."

Because of the legal implications, infringement searching is frequently carried out by an attorney. When the scientist is confronted with an infringement search or, in fact, any patent search, it is important, therefore, that only technical, not legal, decisions be made. The final decision on each suspected claim uncovered in an infringement search must be reserved for the attorney. The best place to carry out an infringement search is in the classified files of the PTO, in spite of the fact that the classification system is designed for anticipation rather than infringement searching. The hierarchical arrangement of the system permits the searching of increasing levels of specificity within each hierarchy while avoiding nonpertinent classes. This leads the inquirer to the collection of patents on which the negative test for infringement is then applied. Unfortunately, this also leads the inquirer to large generic classes that add significantly to the time required to complete the search. Furthermore, there are hidden pitfalls in this approach. An order-of-precedence rule within the classification system may place pertinent patents in an unsuspected class. Other arbitrary rules place further burdens on the searcher. Compositions, for example, are classified according to disclosed use which may not be a restriction to the claims (48). An antioxidant for foods is classified under Class 426, Foods, but may be pertinent to an infringement search on use of the same antioxidant in a lubricating oil. The searcher cannot rely on cross-referencing to compensate for the anticipation emphasis. Under any circumstances, both from the legal and search-strategy standpoints, expert advice should be obtained before undertaking an infringement search.

Validity Searches. When one or more potentially dominating patents are found during the course of an infringement search, a legal opinion can be sought that such dominating claims are invalid. They may be judged invalid on the basis of legal flaws found through a study of the history of prosecution of the application containing such claims or on the basis of anticipating references not uncovered previously. The search conducted to find such anticipating references is called a validity search. It is similar in logic to a novelty search but is more intense. At this point, the stakes are usually higher; for example, the operation of a new plant or marketing of a new product may hinge upon the outcome. Accordingly, the search must be comprehensive, and should include foreign patents as well as the general literature. The uncommon and sometimes hard-to-find sources are frequently important. There is usually an economic incentive to go further and deeper than with any other patent searching. Not only is the search broadened to include many and unusual sources, but the examination of the references themselves is intensified. The pertinent piece of information may be a supplementary or supporting comment found only by careful reading of the text of the full document. Again, as in the case of the novelty search, direct references are not often found, but only close or combination references. It is important, however, to cite these in the search. Even if they cannot be used to invalidate, they may be sufficiently close to put restrictions on the interpretation of the offending claims, thus allowing the potential infringer to circumvent them.

Equivalent Searches. The priority rights provision of the Paris Convention gives rise to the granting of a patent on the same invention in two or more countries. When multiple applications are filed under the priority convention, the priority or first application is referred to in the subsequent documents by application number and filing date. This information allows the tracing of equivalents or patent family members. Files or indexes designed to identify equivalents are used for a number of purposes. Frequently, a chemist or chemical engineer is alerted to a foreign patent of interest. An English-language equivalent (or French or German) should be searched for. The patent-family-member file can save significant time, effort, and translation costs. These files can also be used to follow the patent activities of a competitor. Once the first publication has been detected, much can be learned by watching for later equivalents. Quick alerting may be needed for possible opposition. The publication of equivalents in particular countries may signal manufacturing or marketing plans. If the invention is in a particularly sensitive area, it may be important to plan early for possible conflicts. Use of the patent-family-member file to identify equivalents, thus avoiding duplicate abstracting and/or indexing of the same patent, has saved Derwent and Chemical Abstracts Service considerable time and money over the years.

Both Derwent and INPADOC provide patent-family files. At Derwent an attempt is made to include nonconvention equivalents. These are documents based on applications for the same invention wherein the applications were filed without the benefit of the priority convention. This usually occurs because the applications were not filed within the prescribed time. Although INPADOC covers only convention equivalents, nonconvention family members can frequently be found by reference to their patent-inventor service. *Chemical Abstracts* also provides a patent-concordance file. According to a recent agreement between CA and INPADOC, however, the CA concordance is no longer searchable on-line. The INPADOC patent-family file is considered to be the most complete (49–50). It now covers 47 countries and two regional offices. Derwent covers 24 countries and two regional offices. With respect to recall for the countries covered, Derwent and INPADOC are judged to provide essentially the same efficiency (50). The Derwent service is available on line. Individual patent-family searches on the INPADOC microfiche are provided in the United States by IFI/Plenum Data Company. INPADOC also provides a unique computerized watch service. Any change in the status of a competitor's patent is monitored on a weekly basis. The publication of new equivalents and new document levels are identified by this service. It is practical to monitor hundreds or even thousands of a competitor's active applications in this way.

Business-Oriented Searches.   More and more patent information is being used for business purposes. From the viewpoint of the marketing manager, the following situations call for a study of patent information (51): licensing vs self-commercialization, company acquisitions, competitive analysis, keeping abreast of marketing trends, development of new business and foreign marketing. A similar analysis (52) lists the following uses for patent information in relation to new product management: establishment of the state of the art and trends in technology, identification of companies working in specific areas (including known and potential competitors), identification of personnel working on specific projects, identification of new products, processes, materials, and components, avoidance of duplication of R&D effort, and assistance in marketing strategy.

Patent data may be used in analyzing and forecasting technological trends and the results may be of particular value to marketing managers and research strategists (23). The Office of Technology Assessment and Forecast (OTAF) maintains a bibliographic data base of U.S. patents and provides statistical data analysis in terms of U.S. patent-classification codes, assignees, inventors, application dates, and certain other elements (53). The obvious thrust of this program is to provide information useful in answering business rather than technical questions. A number of patent-information tools have been developed specifically to answer marketing and other business needs. The basic tools are the assignee and inventor indexes provided in the *Annual Index of U.S. Patents* (PTO). However, this index is usually not published until six months or longer after the end of the year and, furthermore, titles are not provided for all entries. The *IFI Assignee Index* also provides quarterly information. The first three issues of the year are available three to four weeks after the end of the quarter, and an annual cumulative edition is available six to eight weeks after the end of the year. Titles are provided for all entries, and patents listed under each assignee are sorted according to classification. Assignee and inventor information with respect to foreign patents is best obtained from the *Chemical Abstracts* and Derwent and INPADOC files. Inventor information in the Derwent files, however, is limited to the on-line system and covers only recent years.

The IFI annual *Patent Intelligence and Technology Report* reports the patent activities for all assignees receiving 10 or more patents in the year. A six-year activity profile is provided for each company along with several unique indexes. Customized intelligence reports may be secured from OTAF, INPADOC, and IFI. In each case, basic bibliographic data (assignee names, inventor names, class codes, etc) stored in data bases are manipulated in a variety of ways to provide statistics, listings, charts, and summaries.

International Aspects

The beginnings of international cooperation in matters of inventions and other intellectual property go back to 1883, with the adoption of the Paris Convention for the Protection of Industrial Property. Since then numerous meetings have been held, and a number of treaties and agreements have been signed in relation to specific aspects of intellectual-property protection. In 1967, the World Intellectual Property Organization (WIPO) was established (54). This body entered into force in 1970 for the purpose of uniting the separate intellectual-property unions that had been established previously or that might be established in the future. More recent developments of importance to patent literature were adoption of the Patent Cooperation Treaty (PCT) in 1970 and the International Patent Classification Agreement in 1971. In December of 1974, WIPO became a specialized agency of the United Nations, and its secretariat is known as the International Bureau. This bureau is the administrative arm of WIPO and has direct responsibility for the activities among the various intellectual property unions. With respect to patents, the International Bureau acts as the administrative coordinator for the Paris, PCT, and IPC Unions.

Tables 15 and 16 summarize patent activity in various countries. Additional up-to-date information is given in refs. 55–56.

The Paris Convention. The Paris Convention for the Protection of Intellectual Property was adopted on March 20, 1883, by 11 states. Since that time, it has been revised six times. By May 1, 1979, the Union had 88 members. The provisions of the convention may be divided into three main sections: equal protection for foreigners, priority rights, and common rules. The equal-protection provision mandates that each member state must grant the same protection to foreigners as it grants to its own nationals. The priority-rights provision allows an applicant for a patent a period of 12 months after filing of the first application to apply for protection in any or all other member states. By filing under the priority convention, the applicant may take advantage of the earliest filing or priority date in any subsequent litigation involving the invention in any of the convention states. In effect, the later-filed applications are regarded as having been filed on the date of the first application. There are important legal advantages in filing under this procedure. From the patent documentation standpoint, priority data have become significant bibliographic information. Their recording is the subject of several important files.

The convention is governed by several common rules. The most important of these, from the patent-literature viewpoint, requires that each country belonging to the union maintain a central office for industrial property and publish a periodic journal listing the owners of the patents granted and abstracts of the patented inventions. Accordingly, the national patent-office journals have, through the years, provided a principal source of patent information.

Table 15. Patent Activities in Various Countries

Country code	Country	International participation[a]: Paris Convention	WIPO	PCT	IPC	EPC	Search authorization	Publication — Examining procedure[b]: Slow 1	Slow 2	Fast 3[c]	Fast 4	Novelty required[d]: Publication	Use	Not patentable[e]: Chemicals	Medicines	Foods
AU	Australia	X	X		X					X		D	D		X	X
AT	Austria	X	X	X	X	X	X		X			C	D	X	X	X
BE	Belgium	X	X		X	X					X	C	D			
BR	Brazil	X	X	X	X					X		C	C	X	X	X
CA	Canada	X	X				X					C	D		X	X
CS	Czechoslovakia	X	X		X				X			C	C	1	1	1
DK	Denmark	X	X	X	X	X				X		C	C		2	2
EP	European Patent Organization	X	X	X	X	X	X			X		C	C			
FI	Finland	X	X		X					X		C	C		2	2
FR	France	X	X	X	X	X				X		C	C		2	2
DD	German Democratic Republic	X	X		X					X		C	D	X	X	X
DE	Federal Republic of Germany	X	X	X	X	X				X		C	C			
HU	Hungary	X	X						X			C	C	X	X	X
IE	Ireland	X	X		X	X			3			C	C		X	X
IL	Israel	X	X		X				X			C	C			
JP	Japan	X	X	X	X		X			X		C	D			
LU	Luxembourg	X	X	X	X	X					X	C	C	X	X	X
NL	Netherlands	X	X	X	X	X				X		C	C			
NZ	New Zealand	X								X		D	D		X	X
NO	Norway	X	X		X	X				X		C	C		2	2
PT	Portugal	X	X		X					X		C	D	X	X	X
RO	Romania	X	X	X						X		C	C	1	1	1
ZA	South Africa	X	X								X	C	C			
ES	Spain	X	X		X						X	C	C		X	X
SE	Sweden	X	X	X	X	X	X			X		C	C			
CH	Switzerland	X	X	X	X	X			X		X	C	C			
GB	United Kingdom	X	X	X	X	X				X		C	C			
US	United States	X	X	X	X		X	X				C	D			
SU	USSR	X	X	X	X		X	X				C	C	1	1	1
WO	WPO											C	C			
YU	Yugoslavia	X	X							X		C	D	X	X	X

[a] WIPO = World Intellectual Property Organization; PCT = Patent Cooperation Treaty; IPC = International Patent Class; EPC = European Patent Convention.

[b] See Figure 2.

[c] With publication of application at 18 months.

[d] D = domestic. C = complete or absolute.

[e] 1 = With respect to exclusive patents. 2 = until end of a transition period.

The Patent Cooperation Treaty (PCT). When protection for an invention is desired in more than one country, a single international application may now be filed under provisions of the Patent Cooperation Treaty of June 19, 1970. This simplifies the application procedure and reduces search and examining efforts. The actual granting of a patent, however, is left to the national offices to which the application has been directed. Each country in which patent protection has been sought may grant or refuse the issuance of a patent, depending upon its own patent laws.

In order to provide a sufficient and consistent means for examining international applications, the PCT provides for the establishment of a number of authorized searching and preliminary examining centers around the world. It has set minimum requirements in terms of document holdings for qualification and has prescribed certain rules and obligations under which such a center must operate. In 1980, by virtue of their having fulfilled the necessary requirements, the patent offices of Austria, the European Patent Organization, Japan, Sweden, the USSR, the United States, and the World Patent Organization have been designated as searching or preliminary examining authorities by WIPO.

ICIREPAT Standards. Several recommendations of the Paris Union Committee for International Cooperation in Information Retrieval Among Patent Offices (ICI-REPAT) for the standardization of patent documents have been adopted and are being administered through the International Bureau. The most significant of these are the provision for a single front page for each document, the creation of a number code for the bibliographic-data elements, the creation of an alphanumerical code for the different kinds of documents, and the creation of alpha codes for the identification of countries and organizations.

The International Patent Classification Agreement. To answer the need for a universal standard system for classifying patented subject matter, the International Patent Classification (IPC) System was devised, and the International Patent Classification Agreement was signed on March 24, 1971. As of May 21, 1979, 27 states were party to it. The system divides technology into eight main sections and some 50,000–60,000 subdivisions. Each member state is obligated to assign appropriate IPC symbols to each of the patent documents it issues. With over 40 countries now applying IPC codes to their patent documents, international searching and examining procedures have been greatly simplified.

Changes in National Procedures Affecting Patent Literature. During the last few years significant changes have occurred in the patenting procedures of France, the FRG, Japan, and the United Kingdom which have affected the patent literature published in these countries.

France. The French Patent Office operated under the Patent Law of July 5, 1844, until January 2, 1968. On that date a new law, effective as of January 1, 1969, was passed. Subsequently several amendments and decrees were enacted. The most recent was an amendment passed on July 13, 1978, with an effective date of July 1, 1979. Both the 1968 law and the 1978 amendment introduced important changes to French patent procedures.

The *Bulletin Official de la Propriété*, the official publication of the French Patent Office, is published weekly in a format that reflects the recent changes. French law offers three types of patent protection: patents that are granted for a term of twenty years from application date; certificates of addition that are granted for the unexpired term of the principal patent; and certificates of utility that are granted for six years from application date. Certificates of utility are granted without examination. Dates on a granted patent include a demandé (application) date, and délivré (grant) date, a publié (publication) date, and, when applicable, a priority date. Special patents on medicaments were abolished by the 1978 amendment.

Because of the need for a transition period between old- and new-law processing, as many as 16 different kinds of French patent documents exist. The procedures for obtaining a French patent have always included an examination for novelty in other countries as well as in France. Prior to the 1978 amendment French patents were unique since they contained no claims. A resume of the main points of the invention appeared at the end of the specification. The current law stipulates that patent applications must contain one or more claims "defining the scope of protection solicited." Publication of patent applications was initiated by the 1968 law that stipulated that all applications filed on or after January 1, 1969, were required to be published eighteen months after filing date or priority date if one existed; that a novelty examination would be conducted only if one was requested within two years of filing; and that failure to request an examination would result in the application being converted to a certificate of utility. Before 1974, complete enactment of the law regarding novelty examinations was not possible since search files at The Hague, the search facility used by the French, were undergoing a transition period. Applicants whose inventions could not be searched can obtain the required search until June 30, 1981.

Federal Republic of Germany. The FRG patents and examined patent applications have been an excellent source of technological information for many years. Both have been published in the *Patentblatt*, the official FRG patent journal, and made available in printed form. In 1968, a new patent law was enacted that extended publication to the unexamined patent application. This law provided that all patent applications filed on or after October 1, 1968, would be published without examination 18 months after application date or priority date if one existed; that applications filed prior to October 1, 1968, would be laid open to public inspection without examination in accord with decisions made by the FRG patent office; that the applicant or a third party could request an examination any time within seven years of application; that any application that survived examination would be published as an examined application; and that opposition to an examined application must be filed within three months from the publication date.

It was further stipulated that the document would retain the number first assigned to the application when it became available as an examined application and later as an issued patent. The terms Offenlegungsschrift (OLS) for unexamined applications, Auslegeschrift (DAS) for examined applications, and Patentschrift (PS) for issued patents were to be used to distinguish the three levels of publication. Numbering of patent applications coming under the 1968 law started with 1,400,000. DAS numbers from 1,000,000 through 1,399,999 had been assigned to some published examined applications prior to passage of the 1968 law; accordingly, this series of numbers has been reserved for applications filed before October 1, 1968.

Basically, the provisions of the 1968 law on publication and opposition have remained unchanged. However, an amendment that changes the duration of a patent and the novelty requirements was passed in 1976. Only applications filed on or after January 1, 1978, are governed by the amended law. Patents and secret patents under government control that were filed before January 1, 1978, have a duration of 18 years beginning on the day following the filing date. Patents and secret patents under government control filed on or after January 1, 1978, have a duration of 20 years beginning

Table 16. Patent Publications in Various Countries

Country code	Title	Official Gazette information					Opposition period	Term of patent
		Periodicity[a]	Bibliographic entries for	Listing order[b]	Title	Abstract		
AU	The Australian Official Journal of Patents, Trade Marks and Designs	W	applications,	applicant	X		informal—between OPI[b] date and acceptance	16 years from filing complete specification
			provisional applications,	applicant	X			
			complete abridgements of accepted specifications	document no.	X	X		
AT	Österreichisches Patentblatt II. Teil	M	examined applications laid open	national class	X		4 months from publication of examined application	18 years from publication of examined application
			granted patents	document no. national class	X			
BE	Recueil des brevets d'invention	M	granted patents	IPC	X	X	none	20 years from application date
BR							90 days from publication of request for examination	15 years from application date
CA	The Patent Office Record	W	granted patents	document no.	X		none	17 years from issue
CS	Věstník úřadu pro vynalezy a objevy	M	applications for inventions	IPC	X		3 months from announcement of acceptance	15 years from application date; unlimited for certificates
			granted author's certificates and granted patents	document no.	X			
DK	Dansk Patenttidende	W	applications accessible to public	IPC	X		3 months from announcement of acceptance	17 years from application date
			examined applications laid open	IPC	X			
EP	European Patent Bulletin	S	granted patents	IPC	X		9 months from grant	20 years from application date
			published applications	IPC	X			
			granted patents	IPC	X			
FI	Patenttidning	M	applications accessible to public	IPC	X		3 months from announcement of	17 years from application date

Table 16 (continued)

Country code	Title	Official Gazette information						Term of patent
		Periodicity^a	Bibliographic entries for	Listing order^b	Title	Abstract	Opposition period	
FR	*Bulletin Officiel de la Propriété Industrielle, "Listes"*	W	applications for patents of invention (first publications)	document no. priority date IPC applicant			third parties may write objections to Patent Office	20 years from application date
			patents of invention (first and only publications)	document no. priority date IPC applicant				
			patents of invention (second publications)	document no.				
	Bulletin Officiel de la Propriété Industrielle, "Abrégés"	W	applications for patents of invention	document no. IPC	X	X		
			patents for invention	document no. IPC	X	X		
DD	*Bekanntmachungen des Amtes für Erfindungs—und Patentwesen*	W	granted economic patents	document no. IPC	X	X	protest by third parties possible	18 years from application date
			granted exclusive patents	document no. IPC	X	X		
DE	*Patentblatt*	W	published unexamined applications	document no. IPC	X	X	3 months from publication of examined application	20 years from day after application date
			published examined applications	IPC	X			
			granted patents	IPC	X			
HU	*Szabadalmi Közlöny és Védjegyértesitő*	M	published patent applications	document no. IPC	X	X	3 months from publication of examined application	20 years from application date
			granted patents	document no. IPC	X			
				IPC	X			
IE	*The Official Journal of Industrial and Commercial Property*	S	applications for patents	applicant	X		3 months from publication of complete specification	16 years from filing complete specification
			complete specifications accepted	document no.	X	X		
IL	*Patents and Designs*	M	applications filed	application no.	X		3 months from publication	20 years from application

Country	Journal	Code	Type of document	Numbering			Opposition period	Term of protection
	Journal							
			examined applications laid open	IPC	X		acceptance	
			granted patents	IPC	X	X		
			applications accepted	document no.		X	of accepted application	
			patents granted	document no.		X		date
JP	*Kokai Tokkyo Koho*	D	published unexamined applications	document no.	X	X	2 months from publication of examined application	15 years from publication date of examined application
	Tokkyo Koho	D	published examined applications	document no.	X			no more than 20 years from application date
LU	*Mémorial Journal Officiel du Grand Duché de Luxembourg Recueil Administratif et Economique*		brevets d'invention	document no.; filing date	X		none	20 years from day after application date
NL	*De Industriële Eigendom, Deel I*	S	published unexamined applications	IPC	X	X	4 months from publication of accepted application	20 years from application date—for patents granted after February 1, 1979
	De Industriële Eigendom, Deel II	M	published examined applications	document no.	X			
NZ	*Patent Office Journal*	M	applications for patents	IPC; application no.	X		3 months from publication of acceptance of complete specification	16 years from filing of complete specification
			complete specifications accepted	document no.	X	X		
NO	*Norsk Tidende for det industrielle Rettsvern, Del I Patenter*	W	applications accessible to public	IPC	X	X	3 months from announcement of acceptance	17 years from application date
			published examined applications	IPC	X			
			granted patents	IPC	X	X		
PT	*Boletin da Propriedade Industrial*		applications for patents	IPC	X	X	3 months from publication of claims in official bulletin	15 years from grant
RO	*Bulletin de informare pentru inventii si marci*	M	patents of invention granted	IPC	X	X	3 months from grant	15 years from application date
			certificates of invention registered	IPC	X			

Table 16 (*continued*)

Country code	Title	Official Gazette information					Opposition period	Term of patent
		Periodicity[a]	Bibliographic entries for	Listing order[b]	Title	Abstract		
ZA	Patent Journal	M	applications for patents	application no.	X		2 months from announcement of acceptance	20 years from date of application
			complete specifications accepted	application no.	X	X		
SU	*Official Bulletin*	F	authors' certificates	document no.	X	X	none	15 years from application, certificates unlimited
			patents of invention	document no.	X	X		
ES	*Boletin oficial de la Propiedad Industrial II Patentes y Modelos de Utilidad*	S	patents granted	document no.	X		none	20 years from issue
SE	*Svensk Patenttidning*	W	applications filed	application no. IPC	X		3 months from announcement of acceptance	20 years from application date
			patent applications accessible to public		X			
			examined applications laid open	IPC	X			
			granted patents	IPC	X			

Code	Publication	Freq[a]	Documents published	Index			OPI	Term of protection
CH	*Schweizerisches Patent- und Markenblatt*	S	patent applications	document no.	X		5 months from publication of examined application	20 years from application date
			granted patents	document no.	X			
GB	*Official Journal (Patents)*	W	applications (1977 Act)	applicant	X		informal—between OPI[b] date and acceptance	20 years from application date
			published applications (1977 Act)	document no.	X			
			division applications (1949 Act)	national class	X			
			complete specifications accepted (1949 Act)	applicant		X		
				document no.				
US	*Official Gazette of the United States Patent and Trademark Office*	W	defensive publications	document no.	X	X	none	17 years from issue
			reissue patents	national class	X	X		
			patents	document no.	X	X		
				national class				
				document no.				
				national class				
WO	*PCT Gazette*	S	published international applications	document no.	X	X		
YU	*Patentni Glasnik*	S	examined applications	IPC	X	X	3 months from publication of examined application	15 years from publication of application
			granted patents	document no.	X			
				IPC				

[a] B = biweekly. F = four-monthly. M = monthly. S = semimonthly. W = weekly.
[b] IPC = International Patent Class; OPI = Opened to Public Inspection.

on the day following the filing date. Patents of addition have a duration of the unexpired term of the main patent regardless of filing date.

Patents filed before January 1, 1978, must have an inventive feature that has not been used in the FRG or described in a printed publication anywhere during the past one hundred years. Patents filed on or after January 1, 1978, must have an inventive feature that is absolutely novel in terms of use, publication, and time period. In both cases, the novelty requirement excludes the six-month period before filing date.

Progression of a patent application from the unexamined version to the issued patent entails three printings: the OLS on yellow or orange paper, the DAS on green paper, and the PS on white paper. It is reported that the FRG plans to eliminate the Auslegeschrift (DAS level of publication) in 1981. Each change in status is published in the *Patentblatt*. However, since examination may be requested over an extended period, applications published for opposition do not appear in a numerical sequence except within a weekly issue of the journal. This makes it extremely difficult to check on a change in status of a particular application.

Japan. Significant changes in the treatment of patents have occurred during the last decade or two. In 1959 the Industrial Property Laws (1959) were enacted. Although Japan had become a party to the Paris Convention in 1899, few changes were made between that time and 1959. The tremendous technological growth and economic development in Japan following World War II soon overwhelmed the laws of 1959. The examination process could no longer keep pace with new applications and by 1970 the average time to process a patent or utility-model application was five years and three months. The number of annual applications filed had increased by 304% for the period 1958 through 1961 (57). As a result of this surge of patent activity, the Patent Law and Utility Model Law (1970) was enacted. Statistics for 1978 provide an indication of recent patent activity in Japan. In that year, 166,092 patent applications were submitted to the Japanese Patent Office. Of these, 24,575 were from foreigners; U.S. applications accounted for 10,316 or 42%. The 1970 revision to the Japanese patent law introduced a procedure for early publication. Under this law, unexamined applications are published 18 months after filing. Request for examination by the applicant or a third party may come any time within seven years of the application date. Examined applications are published a second time; once published, they may be opposed by an adversary if action is taken within two months of publication. This puts a time burden on the western competitor and a priority on fast western-language alerting abstracts. The term of a Japanese patent is 15 years from the publication date but not more than 20 years from the application date. In 1975, revisions were made to the Industrial Property Laws that were particularly significant to the chemical industry. For the first time foods, medicines, and chemical compositions could be patented. The most recent changes in the laws were enacted in 1978 when modifications were made to accommodate the Patent Cooperation Treaty. On October 1, 1978, Japan became a party to this treaty and to the Agreement on the International Patent Classification. At the same time the Japanese Patent Office became an international searching authority and an international preliminary examining authority under the PCT.

Copies of Japanese applications, both unexamined (Kokai) and examined (Kokoku), are printed in the *Japanese Patent Gazette*. This publication is available at the U.S. PTO. It is important to note in referencing Japanese patent documents that both types of applications are given sequential numbers starting each year with number one. The only distinction that shows on the face of the document is that the word Kokai appears on the unexamined application and the word Kokoku on the examined application.

Because of the language barrier, the best way for the non-Japanese to follow Japanese patents is through secondary sources. The INPADOC microfiche services provide complete and timely bibliographic information, and *Chemical Abstracts* and Derwent provide excellent abstract coverage of the Japanese patent literature. English abstracts of unexamined patent applications are also available through the Japanese Institute of Invention and Innovation (JIII). Although the JIII abstracts go back only to 1976, their importance will grow as the number of years they cover increases. Copies of Japanese patent documents may be obtained directly from the Japanese Patent Office or through several commercial services discussed above.

United Kingdom.   English patent law dates back to the 1600s (58). As in other countries, many changes were made in the course of the industrial revolution. Following World War II, the Patent Act of 1949 governed patent practice with modifications in 1967. In 1977 significant changes were made to bring UK law in line with the terms of the European Patent Convention of 1973. The United Kingdom is a member of the Paris Convention, the World Intellectual Property Organization, the Patent Cooperation Treaty, the International Patent Classification Agreement and the European Patent Convention (EPC).

Because of the various changes made in the law during the last 20 years, five kinds of documents may issue: Patents based on applications dated before June 1, 1978 (old law); patents of addition based on applications dated before June 1, 1978 (old law); patents based on applications dated June 1, 1978 or later (new law); patents based on European patent applications; and patents based on international patent applications. The possibility of old-law and new-law documents issuing exists likewise in France, the FRG, and Japan where recent changes also admit to the publication of patent documents under two procedures. Applications for patents of addition and provisional applications are no longer permitted in the United Kingdom.

Under the new UK law, unexamined applications are usually published 18 months (frequently several weeks later, in practice) after the filing or priority date. However, an applicant may request earlier publication in order to preempt a possible conflicting application elsewhere. Following filing, a preliminary search is made and a report sent to the applicant, who has six months from the publication date to request a substantive examination. At this time the application is examined for novelty and patentability. Upon acceptance, usually two and one-half years from application, notice is published in the *Official Journal* (*Patents*). Printed copies of the accepted specification become available two to three months later. The application is given a five-digit number preceded by the year (YYNNNNN). The first publication is assigned a seven-digit number (currently in the 2×10^6 series) followed by the letter A. Upon grant, the accepted specification is reprinted and given the same seven-digit number followed by the letter B. Although formal opposition procedures are no longer applicable, the final grant of a patent may be refused any time between the time of first publication and grant, if an anticipatory reference is brought to the attention of the patent office during this period.

The UK patent gazette, the *Official Journal* (*Patents*), is published weekly on Wednesdays. It includes a list of new applications arranged alphabetically by applicant.

Filing information and a brief title are also provided. In addition, the gazette lists unexamined applications (OPI 18 months) arranged in UK classification order and accepted patents in patent number order, with cross indexes based on application number, applicant name, British classification number, and abridgement-allotment number. The UK Patent Office also prepares and publishes classified *Abridgements of Specifications*. Since 1931 a group volume has been published every 20,000 patent numbers, with the various classes divided into 40 groups up to specification number 600,000, then into 44 groups up to specification number 940,000, after which they are consolidated into 25 groups. From specification number 1,000,001, each volume, with its related indexes, covers 25,000 numbers. Each volume contains guides to classes and groups along with detailed illustrated abridgements arranged by patent number. Each group volume has a corresponding index by subject and name; for the whole number series, a Group Allotment Index appears within two months of the last specification, with the information where the abridgement of a given patent number can be found (also published weekly in the Official Journal). Entrance by name is by the index to names of applicants mentioned above. Although the pamphlet version of the abridgements is issued within seven days of publication of the specification, the regular, cumulative series requires four to eight months for completion.

Outlook

The concept of governments providing protection to inventors dates back to the 13th century. The first record of a granted patent was that by the Republic of Florence in 1421 for a barge fitted with hoisting gear to load and unload marble (58). In the United States the power "To promote the progress of science and useful arts . . ." is granted through Article 1, section 8 of the Constitution. Some 130 countries throughout the world currently operate patent offices for the purpose of protecting inventors and encouraging disclosure. Regardless of the diversity in detailed practice of the many patent systems of the world, they all share the same fundamental goal. That goal is to provide an incentive to each inventor to disclose his or her findings for the long-term benefit of society rather than to attempt to profit from the invention in secret (59). This drive to promote disclosure of inventions continues unabated. Over 10^6 patent documents will be published in 1981. The public's ability to cope with this flood of information is severely taxed. Publication has been extended beyond present ability to make effective use of the record (60). However, a new era of more effective electronic communications is beginning to emerge as a useful alternative to hard-copy processing. Dissemination of patent information is handled, at least partially, by electronic means. During the decade ahead this application will certainly grow and the ability to keep pace with the ever-expanding volume of patent data will be enhanced by it greatly.

BIBLIOGRAPHY

"Patent Literature" in *ECT* 1st ed., Vol. 9, pp. 890–897; "Patent Literature" in *ECT* 2nd ed., Vol. 14, pp. 583–635, by Errett S. Turner, Bell Telephone Laboratories, Inc.

1. B. F. M. Helliwell, *Inf. Sci.* **8,** 117 (1974).
2. J. F. Terapane, *Chemtech*, 272 (May 1978).
3. H. C. Lynfield, *J. Pat. Off. Soc.* **47,** 374 (1965).
4. M. M. Boguslavsky and co-workers in I. Y. Morozov, ed., *Inventor's Certificate as a Form of Legal Protection of Inventions*, The CNIIPI of the State Committee of the USSR for Inventions and Discoveries, Moscow, USSR, 1978, p. 6.
5. B. A. Ringer and K. M. Mott in R. Calvert ed., *The Encyclopedia of Patent Practice and Invention Management*, Reinhold Publishing Corp., New York, 1964, pp. 198–203.
6. H. C. Robb, Jr. in R. Calvert, ed., *The Encyclopedia of Patent Practice and Invention Management*, Reinhold Publishing Corp., New York, 1964, pp. 641–655.
7. *International Table to Patents*, *Designs, and Trade Marks*, 69th ed., H. Scheer-Verlag, Hürth-Efferen/Köln, FRG, 1979 (wall chart, 2 parts).
8. L. J. Robbins and J. H. Handelman in R. Calvert, ed., *The Encyclopedia of Patent Practice and Invention Management*, Reinhold Publishing Corp., New York, 1964, pp. 25–30.
9. M. M. Duffey, *J. Chem. Inf. Comput. Sci.* **17,** 126 (1977).
10. M. G. E. Luzzati, *Trans. Chart. Inst. Pat. Agents* **80,** C33 (1961–1962).
11. R. K. Summit, *Remote Information Retrieval Facility*, N-07-68-1, NASA CR-1318, Lockheed Palo Alto Research Laboratory, Palo Alto, Calif.
12. P. L. Dedert and K. E. Shenton, "A Comparison of Using CA Search on SDC and Lockheed Search Service," *presented at the National Online Information Meeting, New York*, Mar. 25–27, 1980.
13. R. G. Smith, L. P. Anderson, and S. K. Jackson, *J. Chem. Inf. Comput. Sci.* **17,** 148 (1977).
14. S. M. Kaback, *Online*, 16 (Jan. 1978).
15. R. J. Rowett, Jr., *Chemtech*, 348 (June 1979).
16. *CLAIMS™ File 25*, DIALOG Online Retrieval System, database supplier IFI/Plenum Data Company, Arlington, Va., Mar. 18, 1980.
17. J. T. Maynard, *Chem. Inf. Comput. Sci.* **17,** 136 (1977).
18. *Chem. Eng. News*, 50 (Mar. 24, 1980).
19. H. Hyams, *WIPO Moscow-Symposium*, reprint of lecture, Derwent Publication Ltd., London, UK, 1974.
20. S. M. Kaback, *Chemtech*, 172 (Mar. 1980).
21. S. M. Kaback, *J. Chem. Inf. Comput. Sci.* **20,** 1 (1980).
22. *CHRONOLOG™*, Monthly Newsletter, Lockheed Dialog Information Retrieval Service, Palo Alto, Calif., Mar. 1980, p. 3.
23. H. M. Allcock and J. W. Lotz, *J. Chem. Inf. Comput. Sci.* **18,** 65 (1978).
24. H. M. Allcock and J. W. Lotz, *Chemtech*, 532 (Sept. 1978).
25. P. W. Howerton, "Automation and Scientific Communication," *Amer. Doc. Inst., 26th annual meeting, Chicago*, Oct. 6–11, 1963, Short papers Part 2, p. 155.
26. P. W. Howerton, *J. Chem. Doc.* **4,** 232 (1964).
27. P. T. O'Leary, J. M. Cattley, J. E. Moore, and D. G. Banks, *J. Chem. Doc.* **5,** 233 (1965).
28. J. M. Cattley, J. E. Moore, D. G. Banks, and P. T. O'Leary, *J. Chem. Doc.* **6,** 15 (1966).
29. J. M. Cattley, T. A. Reif, J. E. Moore, D. G. Banks, and P. T. O'Leary, *Chem. Eng. Prog.* **62,** 91 (1966).
30. L. E. Rasmussen and J. G. Van Oot, *J. Chem. Doc.* **9,** 201 (1969).
31. M. Z. Balent and J. M. Emberger, *J. Chem. Inf. Comput. Sci.* **15,** 100 (1975).
32. M. Z. Balent and J. W. Lotz, *J. Chem. Inf. Comput. Sci.* **19,** 80 (1979).
33. K. M. Donovan and B. B. Wilhide, *J. Chem. Inf. Comput. Sci.* **17,** 139 (1977).
34. *Ind. Property*, 290 (1973).
35. *INPADOC General Information*, WIPO Publication 426 (EFG), World Intellectual Property Organization, Geneva, Switz., 1979, p. 5.
36. *CHRONOLOG™*, Monthly Newsletter, Lockheed DIALOG Information Retrieval Service, Palo Alto, Calif., Jan. 1980, p. 3.
37. Y. Kameyama, Private communication, Japan Institute of Invention and Innovation, Tokyo.
38. S. M. Kaback, K. Landsberg, and A. Girard, *Database* 1(2), 46 (1978).
39. L. Rogalski, *J. Chem. Inf. Comput. Sci.* **18,** 9 (1978).
40. R. E. Maizell, *How to Find Chemical Information*, John Wiley & Sons, Inc., New York, 1979, p. 127.
41. Ref. 40, p. 130.

42. F. W. Lancaster, *Vocabulary Control for Information Retrieval*, Information Resources Press, Washington, D.C., 1972, p. 107.
43. J. T. Maynard, "How to Read a Patent," in *Understanding Chemical Patents*, American Chemical Society, Washington, D.C., 1978, p. 7.
44. S. M. Kaback, *J. Chem. Inf. Comput. Sci.* **17**, 143 (1977).
45. *U.S. Pat. Q.* **58**, 74 (1943).
46. A. H. Seidel, "The Client's Invention and Its Patenting," in *The Practical Lawyer*, ALI-ABA, Philadelphia, Pa., Vol. 1, No. 1, 1955, pp 53–66.
47. A. K. Berle and L. S. De Comp, "Domination by Prior Patents," in *Inventions and Their Management*, Laurel Publishing Co., New York, 1954, pp. 225–235.
48. I. R. Lady, J. R. Leclair, I. J. Rotkin, and H. S. Vincent, *Development and Use of Patent Classification Systems*, U.S. Department of Commerce/Patent Office, Washington, D.C., 1966, p. 14.
49. Ref. 40, p. 152.
50. T. M. Johns and co-workers, *J. Chem. Inf. Comput. Sci.* **19**, 241 (1979).
51. N. H. Giragosian, *J. Chem. Inf. Comput. Sci.* **18**, 121 (1978).
52. M. S. White, *Int. J. New Prod. Manage.* **2**, 67 (1979).
53. *Technology Assessment & Forecast*, Program Brochure, Office of Technology Assessment, and Forecast, U.S. Patent and Trademark Office, Washington, D.C.
54. *World Intellectual Property Organization*, General Information, WIPO Publication No. 400 (E), Geneva, Switz., 1979, p. 5.
55. *Manual for the Handling of Applications for Patents, Designs and Trademarks Throughout the World*, Octrooibureau Los en Stigter, Amsterdam, 1978 (looseleaf).
56. *Patents Throughout the World*, Digest of Patent Laws (looseleaf), Trade Activities, Inc., New York, 1978.
57. *Japanese Patent Office Annual Report 1978*, English ed., Tokyo, p. 50.
58. H. Skolnik, *J. Chem. Inf. Comput. Sci.* **17**, 119 (1977).
59. Ref. 43, p. 5.
60. V. Bush, *Atl. Mon.* **176**(1), 101 (July 1945).

General References

P. D. Rosenberg, *Patent Law Fundamentals*, Clark Boardman Co., Ltd., New York.
International Patent Applications According to PCT, H. Scheer-Verlag, Hürth-Efferen/Köln, FRG, 1978.
J. Fichte, H. Marchart, and G. Quarda, *Austria's Patent Information Services for Developing Countries*, Austrian Federal Ministry for Science and Research, UNCSTD, Vienna, 1979.
G. Kalugin and F. Goncharenko, *Patenting of Foreign Inventions in the USSR (Practical Guide)*, The USSR Chamber of Commerce and Industry, Patent Department, Moscow, USSR, 1975.
International Patent Classification (manual), 3rd ed., WIPO, Carl Heymanns Verlag KG, Munich, FRG, 1979.
World Patent Information, K. G. Sour, Munich, FRG, 1980, quarterly.

JOHN W. LOTZ
IFI/Plenum Data Company